CAMBRIDGE LIBRARY COLLECTION

Books of enduring scholarly value

Monographs of the Palaeontographical Society

The Palaeontographical Society was established in 1847, and is the oldest Society devoted to study of palaeontology worldwide. Its primary role is to promote the description and illustration of the British fossil flora and fauna, via publication of an authoritative monograph series. These monographs cover a wide range of taxonomic groups, from microfossils, trilobites and ammonites through to Coal Measure plants, mammals and reptiles, and from all ages from Cambrian to Pleistocene. They form a benchmark for understanding the past life of the British Isles and many include the original descriptions of numerous key species. The first monograph (on the Crag Mollusca) was published in March 1848 and the Society still continues this work today. Notable authors in the series include Charles Darwin (fossil barnacles) and Richard Owen (dinosaurs and other extinct reptiles). Beginning in 2014, the Cambridge Library Collection and the Society are collaborating to reissue the earlier publications, focusing on monographs completed between 1848 and 1918.

A Monograph of British Belemnitidae

The nephew of William Smith, John Phillips (1800–1874) was also an influential geologist. Professor of Geology at Oxford, and in part founder of the Oxford University Museum, he notably proposed the three main eras: the Palaeozoic, Mesozoic and Cenozoic. Originally published between 1865 and 1870, this monograph was the first to attempt a review of British belemnites, and was influenced in part by the author's youthful studies of the Yorkshire Jurassic. During the mid-nineteenth century there were many developments in the study of this important fossil group, with the discovery of soft parts, and documentation of diverse faunas across Europe. Phillips' work complemented these advances. Though he died before completing the Cretaceous section, the monograph covers many of the most important species (though, as was common at the time, restricted to the genus *Belemnites* in the broad sense). It remains a starting point for any study of British belemnites.

A Monograph of
British Belemnitidae

John Phillips

CAMBRIDGE
UNIVERSITY PRESS

CAMBRIDGE
UNIVERSITY PRESS

University Printing House, Cambridge, CB2 8BS, United Kingdom

Cambridge University Press is part of the University of Cambridge.

It furthers the University's mission by disseminating knowledge in the pursuit of
education, learning and research at the highest international levels of excellence.

www.cambridge.org
Information on this title: www.cambridge.org/9781108081238

© in this compilation Cambridge University Press 2015

This edition first published 1865–1909
This digitally printed version 2015

ISBN 978-1-108-08123-8 Paperback

This book reproduces the text of the original edition. The content and language reflect
the beliefs, practices and terminology of their time, and have not been updated.

Cambridge University Press wishes to make clear that the book, unless originally published
by Cambridge, is not being republished by, in association or collaboration with,
or with the endorsement or approval of, the original publisher or its successors in title.

The original edition of this book contains a number of oversize plates
which it has not been possible to reproduce to scale in this edition.
They can be found online at www.cambridge.org/9781108081238

A MONOGRAPH

OF

BRITISH BELEMNITIDÆ:

JURASSIC.

BY

JOHN PHILLIPS,

M.A.OXON., LL.D.DUBLIN, LL.D.CAMBRIDGE, D.C.L.OXON., F.R.S., F.G.S., ETC.; PROFESSOR OF GEOLOGY
IN THE UNIVERSITY OF OXFORD.

LONDON:
PRINTED FOR THE PALÆONTOGRAPHICAL SOCIETY.
1865—1909.

PRINTED BY ADLARD AND SON, LONDON AND DORKING.

INDEX

TO SPECIFIC NAMES OF BRITISH JURASSIC BELEMNITES DESCRIBED OR DISCUSSED.

ADLARD AND SON, IMPR., LONDON AND DORKING.

THE

PALÆONTOGRAPHICAL SOCIETY.

INSTITUTED MDCCCXLVII.

VOLUME FOR 1909.

LONDON:

MDCCCCIX.

A MONOGRAPH OF BRITISH BELEMNITIDÆ: JURASSIC.

ORDER OF BINDING AND DATES OF PUBLICATION.

PAGES	PLATES	ISSUED IN VOL. FOR YEAR	PUBLISHED
General Title-page and Index	—	1909	December, 1909
1—28	—	1863	June, 1865
29—52	I—VII	1864	April, 1866
53—88	VIII—XX	1866	June, 1867
87—108	XXI—XXVII	1868	February, 1869
109—128	XXVIII—XXXVI	1869	January, 1870

THE

PALÆONTOGRAPHICAL SOCIETY

INSTITUTED MDCCCXLVII.

LONDON:

MDCCCLXV.

A MONOGRAPH

OF

BRITISH BELEMNITIDÆ.

BY

JOHN PHILLIPS,

M.A. OXON., LL.D. DUBLIN, F.R.S., F.G.S., ETC.,
PROFESSOR OF GEOLOGY IN THE UNIVERSITY OF OXFORD.

LONDON:
PRINTED FOR THE PALÆONTOGRAPHICAL SOCIETY.
1865.

J. E. ADLARD, PRINTER, BARTHOLOMEW CLOSE.

BRITISH BELEMNITES.

HISTORICAL NOTICES.

BELEMNITES were first so named in Germany, in the celebrated work of Georgius Agricola, of date 1546, and described by him among the 'Figured Stones' which then began to attract attention in Europe. In 1677 they were noticed in England by Dr. Plot ('Natural History of Oxfordshire'), and in 1678, by an equally famous pioneer of natural history, Dr. Lister.[1] From this time the well-known controversy regarding their nature—whether they were mere stones, crystals, horns, or shells—was stoutly maintained by Grew,[2] 1681, Woodward,[3] 1695, Lhwyd,[4] 1699, and others in England, and by many foreign naturalists, until the general progress of zoology and geology left no room for doubt as to their affinity with the shelly supports of other better-known Cephalopoda.

The description of species of British Belemnites begins with Dr. Plot, 1677, the first Keeper of the Ashmolean Museum in Oxford, to whom I am therefore especially bound to render due justice. It is easy to recognise in his writings and figures the large species of Belemnite, called *Belemnites abbreviatus* by a later writer, which occurs in the Coralline Oolite of Headington, near Oxford;[5] *B. elongatus*, Miller, is probably one of the two species from the Upper Lias of Great Rolwright, Oxon.; and the deeply grooved species (*B. sulcatus*, Mill.), is common near Oxford. ('Natural History of Oxfordshire,' pl. iii,—3, 4, 5, 6.)

[1] 'Historiæ Animalium Angliæ, tres tractatus,' 1678, 4to.
[2] 'Catalogue of Rarities,' &c., 1681, fol.
[3] 'Essay towards a Natural History of the Earth,' 1695, 8vo.
[4] 'Lithophylacii Brittannici, Ichnographia,' 1699, 8vo.
[5] "If vehemently rubbed, the only one amongst all that I have that, like amber, takes up straws and some other light bodies." ('History of Oxfordshire,' p. 94.)

Lister follows, in 1678, with special notice of the small mucronate fossil, which he called *B. minimus*, common in the "Red Chalk" of Yorkshire, the county in which he resided. ('Hist. Anim. Angliæ.') He says it is found in all the cliffs as you ascend the Wolds, for above a hundred miles in compass, at Speeton, Londesborough, and Caistor, but always in a red ferruginous earth. This remarkable observation shows how near were the able inquirers of the seventeenth century to the discovery which was made a century later by W. Smith in another part of England.

Lhwyd, in 1699, gave descriptions of no less than eighty-two specimens of Belemnites, with their localities, and figures of two guards, and one phragmocone. The localities indicated show that this diligent man had collected from most of the strata of England, and it is interesting to read his remark that in his native Wales he nowhere found a single Belemnite. Except in the Lias of the southern coast of Wales, no one will be likely to find any now in the otherwise rich principality. Few of the specimens collected by Plot and Lhwyd are now to be found in the Oxford Museum.

The Coralline Oolite and Calcareous Grit yielded to Lhwyd "Belemnites maximus oxyrhynchus," and the large Belemnites, known as *B. abbreviatus*, Mill., from the quarries at Cowley, Bullington, Stansford, Marsham, Garford, Basisleigh—all within easy reach of Oxford. A subfusiform small Belemnite, possibly the young of the large species already named, is quoted from Cowley and Marsham. One locality is given for a large cylindrical species, which is remarkable :—"In medio saxo invenimus, ad collem Garsingtonensem." He also quotes a small specimen from the stone quarries of Thame, and other very small examples. These are the only statements known to me of any Belemnite being found in the Portland Rocks of England.

From the Oxford Clay by the Cherwell we have the young form of *B. sulcatus*, Mill., also recognised at Apsley, in Bedfordshire. Murcot (now Moorcot) on the edge of Otmoor, produced a fine example of *B. tornatilis*, seven inches long. Huntingdon, and Pyrton, in Wiltshire—"e puteo carbonibus eruendis destinato"—are also quoted for Belemnites of this age.

The Bath Oolites, in their various stages, yielded several examples. "Stunsfield," as it was then written, is cited for three caniculate forms, which include the two now generally known as *B. Bessinus* and *B. fusiformis*. At Barrington, Kidlington, and Witney, localities visited near Oxford, examples occurred, but the species are untraceable.

We recognise fossils of the Marlstone from Alderley and Wootton-under-Edge, in Gloucestershire, but cannot determine the species.

Lias Belemnites were collected at Boughton and Marston Trussel in Northamptonshire, Whitton-on-Humber, Pyrton Passage on Severn, and Radstoke, in Somerset, but the descriptions are insufficient for identification.

Chalk Belemnites are mentioned in Kent and Norfolk, but special attention was not directed to them.

Morton, the author of the excellent 'Natural History of Northamptonshire,' in 1712, mentions, but with no critical attention, Belemnites corresponding to the great types already familiar to the English collectors. Fusiform Belemnites ("Belemnites ari pistillum referens") of Lhwyd, with a ventral canal, are mentioned in Upper Lias Clay at Marston Trussel and Oxendon, and in Stone at Crick. It is easy to recognise *B. elongatus* of Lias, and *B. sulcatus* of Oxford Clay, among the contents of Morton's drawers, and it is evident that he looked curiously and carefully at those objects; but he gives no figures, a somewhat remarkable exception to his general practice, which is to present a great number of very tolerable illustrations of invertebral fossils.

Woodward.—Belemnites were much studied by Dr. John Woodward, not merely as to the true nature of these fossils, in which he was signally wrong, but as to the diversity of their forms, and the variety of accidental circumstances by which they are accompanied. In his 'Method of Fossils' (1728) he speaks of "the *Belemnites fusiformis*" of J. Bauhin, another of a different form pointed at each end, and of a "conic" Belemnite. In the 'Natural History of Fossils' (1729) he gives several particulars and a good deal of general information about the conic, fusiform, and bicuspid Belemnites, the two former having one 'chap or seam' running down one side, the latter no such seam. In this his latest notice of Belemnites we find him at last willing to correct his former notions, and to admit that they might be of "animal, and not of mineral nature, as he had ever taken them to be" (p. 104). The species of Belemnites thus defined may be in some degree ascertained by the reference to localities. Thus, the bicuspid sorts, being quoted from the chalk-pits of Northfleet, Greenhithe, and Croydon, are easily recognised as the incomplete fossils called Actinocamax by Miller, now included in the genus Belemnitella. The fusiform Belemnites include the well-known "Stunsfield" species (also quoted from Hannington, in Wiltshire), and one of similar general shape but different in construction from "Spitten," in Yorkshire, where there are "great numbers of them in blue clay in a large cliff." Another locality given is a tile-clay-pit, near Thurnham, three miles from Maidstone, in Kent; this was a species of the Gault, probably that now called *B. minimus*. One other locality is given for a fusiform Belemnite, viz., a quarry half a mile west of Clipston, Northamptonshire; it probably refers to a species in the Marlstone, such as *B. clavatus*, Blainville.

In his group of conic Belemnites Woodward placed the larger portion of the specimens which had come under his notice. Most of them had the "chap or seam running lengthways of the surface," which we find in many Oolitic Belemnites, or else the narrow canal which belongs to the Cretaceous Belemnitellæ.

Belonging to this latter group were specimens from Greenhithe and Northfleet, in Kent, which the author suspected to be of the same kinds as the bicuspid sorts already noticed[1]. The large *Belemnites abbreviatus* of the Coralline Oolite was received by Woodward from Cowley Common, near Oxford, with vermiculi (Serpulæ) and small oysters adherent. An allied species occurred at Stowell, in Gloucestershire, in the Oolitic zone of

---

[1] At present we should refer those incomplete "bicuspid" to these more perfect "conic" examples.

Dundry. The Bath Oolite series also yielded him specimens corresponding or allied to *B. sulcatus* of Miller, from Sherburn, Birliphill, Farmington, Yanworth, Northleach, Colne Allens, in Gloucestershire; and Barrington, in Oxfordshire. The Marlstone of Clipston, in Northamptonshire, furnished specimens probably of *B. paxillosus*.

From the Lias of Boughton and Ashley, in Northamptonshire, joints probably of the phragmocone of *B. elongatus*, Sow.; and from the Upper Lias (Alum Shale) of Whitby several specimens, probably of *B. vulgaris*, Young and Bird.

One remarkable locality, Silverton, Devonshire, is quoted for several specimens of a conic Belemnite, with "a seam or sulcus running down one side of this body for the whole length of it." This obviously describes a canaliculate species of the Inferior Oolite; but Silverton is situated in the Trias. Perhaps we ought to read Silverston, in Northamptonshire. On the whole, it may be concluded that the following species were handled by Dr. Woodward:

CHALK	Belemnitella mucronata, *Schl.* .	. Kent.
,, ...	— plena, *Blainv.* .	. Kent.
GAULT . . .	Belemnites minimus, *Mill.* .	. Kent.
SPEETON CLAY . .	— jaculum, *Ph.* .	. Yorkshire.
OXFORD OOLITE GROUP .	— abbreviatus, *Mill.*	. Oxfordshire.
BATH OOLITE GROUP .	— sulcatus, *Mill.* .	. Gloucestershire.
,, .	— fusiformis, *Mill.* .	. Oxfordshire.
LIAS	— clavatus, *Bl.* .	. Northamptonshire.
,, ...	— vulgaris, *Y.* and *B.* .	. Yorkshire.
,, ...	— elongatus, *Mill.* .	. Northamptonshire.
,, ...	— paxillosus, *Schl.* .	. Northamptonshire.

In 1764 Mr. Joshua Platt, a patient and not unsuccessful explorer of the fossils of Stonesfield, near Oxford, communicated to the Royal Society observations on the structure of Belemnites,[1] which contain the sound opinion that the Belemnite was a shell formed, as the hard parts of mollusca are, by deposition from a secreting surface.

In the elegant work of Mr. Parkinson, 1804, Belemnites appear among the "organic remains of a former world," and the terms employed by Woodward retain their place. The "conic" Belemnite, the "cylindrical" Belemnite, and the "fusiform" Belemnite of Stonesfield, the large, nearly round fossil of the Oxford Clay, and the mucronate form of the Chalk, are represented, but not critically distinguished.

In 1823 the Geological Society of London received from Mr. J. S. Miller, a native of

---

[1] 'Phil. Trans.,' liv, p. 38. Mr. Platt was the "discoverer" of the Stonesfield mammals, though, perhaps, he may not have known the full value of the lower jaw of *Amphitherium Broderipii* (Owen), which formed part of the collection furnished by him to an ancestor of my late friend, the Rev. C. Sykes, of Rooss, who, at my request, gave the specimen to the Yorkshire Phil. Society. (See Owen, 'Brit. Mammal.,' p. 58.)

Dantzic, residing at Bristol, and charged with the custody of the Museum of the Institution there, 'Observations on Belemnites,' in which twelve species were named, described, and figured. Notwithstanding some errors, which may be partly typographical (as when Plot, who wrote in 1677, is represented as communicating a paper to the 'Phil. Trans.' in 1764, the true author being Joshua Platt), this is a treatise of much value, and it made a first and important step in the right direction. He describes and figures the following species, mostly named by himself.

1. *Belemnites abbreviatus.* Pl. VII, figs. 9, 10. From Weymouth and Dundry. Inferior Oolite. The specimen figured is in the Museum at Bristol, and appears really to have been derived from the Coralline Oolite, in which the species is common. It has not been found in the Inferior Oolite at Dundry or elsewhere. The description is inadequate, and other species have since been ranked under this name.

2. *Belemnites aduncatus.* Pl. VIII, figs. 6, 7, 8. From Weymouth and Lyme, in Lias. The former locality is incorrect. It is difficult to identify this species, though specimens occur in Upper Lias which somewhat resemble this figure of Miller.

3. *Belemnites sulcatus.* Pl. VIII, figs. 3, 4, 5. From Dundry, near Oxford. Inferior Oolite. There is more than typographical confusion here. Figs. 3 and 4 are taken from Dundry specimens, and belong to a species found in the Inferior Oolite. Fig. 5 is from an Oxford specimen of a different type, out of the clay of that name. Mistakes in regard to this name are very common.

4. *Belemnites elongatus.* Pl. VII, figs. 6, 7, 8. From Lyme, in Lias. Imperfectly defined, so that different species have been since called by this name.

5. *Belemnites longissimus.* Pl. VIII, figs. 1, 2. From Lyme, in Lias.

6. *Belemnites acutus.* Pl. VIII, fig. 9. No locality ;[1] but reference is made with doubt to Lhwyd, fig. 1683, which represents a fossil from Merston, in Northampton-shire. Much confusion in regard to the application of this name by Blainville and later writers.

7. *Belemnites tripartitus,* Schlottheim. Pl. VIII, figs. 10, 11, 12, 13. No locality given. Not easily identified, among several cognate forms.

8. *Belemnites ellipticus.* Pl. VIII, figs. 14, 15, 16, 17. From Dundry. Inferior Oolite. By many writers identified with *B. giganteus.*

9. *Belemnites electrinus.* Pl. VIII, figs. 18, 19, 20, 21. Salisbury, Brighton, Lewes, Chalk. Some foreign localities are given in error, the Baltic species being different from that of Maestricht.

10. *Belemnites fusiformis.* Pl. VIII, fig. 22; Pl. IX, figs. 5—7. Stonesfield.

11. *Belemnites minimus.* Pl. IX, fig. 6. Folkstone, Ringmer, &c. Gault.

[1] See on a future page the localities now admitted.

Mr. Miller also describes, under the name of *Actinocamax verus*, the incomplete guard of a twelfth Belemnite, from which the part containing the "alveolus" had been removed by decomposition of the nacreous laminæ. Before his death this was made clear to the ingenious author, to whom we are indebted for a still more valuable contribution to palæontology, viz., the essay on 'Crinoïdea.'

For the greatest addition ever made to British fossil conchology we are indebted to James Sowerby, who in 1812 commenced, and James De Carle Sowerby, who continued the labour of engraving the countless mollusca of the strata of the British Isles.

It is only in the sixth volume, and towards the end of that volume, that figures of Belemnites occur. They relate to species which had been previously described, some in English and others in foreign works, and include fossils of the Lias, Oolite, and Chalk.

From the Lias.—*B. pistilliformis*, t. 589, f. 3. *B. penicillatus*, t. 590, f. 5, 6. *B. elongatus*, t. 590, f. 1. *B. acutus*, t. 590, f. 7, 8, 10.

From the Oolites.—*B. abbreviatus*, t. 590, f. 2. *B. compressus*, t. 590, f. 4 (Scarborough). (*B. ellipticus*, mentioned p. 182. *B. gigas*, mentioned p. 182.)

From the Chalk, Greensand, and Gault.—*Belemnites granulatus*, t. 600, f. 3, 5. *B. lanceolatus*, t. 600, f. 8, 9. *B. attenuatus*, t. 589, f. 2. *B. minimus*, t. 589, f. 1. *B. mucronatus*, t. 600, f. 1, 2, 4, 6, 7. (Fig. 1 is referred by Sharpe to *B. lanceolatus*, Schl.)

Beloptera is also noticed, and three species from the Cænozoic series are figured; one, *B. anomala*, from Highgate; the others, *B. belemnitoidea*, and *B. sepioidea*, from France.

W. Smith, in his works entitled 'Stratigraphical System of Organized Fossils' (1817) and 'Strata identified by Organic Remains' (1816 and following years) notices some of the Belemnites in his large collection, now placed in the British Museum.

Among the fossils selected for identifying the Upper Chalk he places *Belemnites mucronatus;* to the Gault, or "Micaceous Brickearth," he assigns *Belemnites minimus;* and to the Oxford or 'Clunch' Clay, the Belemnite which has been since called *Owenii*, and has received other designations. The work was never completed.

In the 'Stratigraphical System' he gives two Belemnites from the Crag, no doubt drifted; one, a siliceous cast of alveolus from the Chalk. Two Belemnites from the Chalk, one from the Upper Greensand, one from the Gault, one from the Kimmeridge Clay, one from the Coralline Oolite, two from the Oxford Clay, one from the Kelloways Rock, one from the Fuller's Earth Rock, two from the Inferior Oolite, one from the sand and sandstone below, four from the Marlstone, with which rock the publication ceased. The species are mostly recognisable, except some of those in the Marlstone; among them may be enumerated the following:

CHALK	Belemnites mucronatus, *Sow.*	
GREENSAND	— plenus ? *Blainv.*	
GAULT	— minimus, *Sow.*	
KIMMERIDGE CLAY . . .	— excentricus, *Blainv.*	
CORALLINE OOLITE . . .	— abbreviatus, *Miller.*	

OXFORD CLAY. ⎫ KELLOWAYS ROCK ⎭ . .	Belemnites Owenii, *Pratt.*
FULLER'S EARTH ROCK ⎫ INFERIOR OOLITE . ⎭ .	— canaliculatus, *Schl.*
MARLSTONE	— paxillosus, *Schl.*
	— elongatus, *Mill.*

In 1822 the Rev. G. Young and Mr. J. Bird issued the first edition of their 'Geological Survey of the Yorkshire Coast.' Among the objects noticed and figured are a few Belemnites:

B. vulgaris, common in the Alum Shale (Upper Lias).
B. excentralis, said to occur in the Oolite (of Malton), the Upper Shale (Speeton), and the Chalk.
B. fusiformis, from the Speeton Shale.
B. tubularis, from the Alum Shale.

In the second edition (1828) one more species, called *B. compressus*, is noticed, and somewhat strangely figured. The names were given without reference to any works but those of Sowerby and Miller, and the species here marked *fusiformis* and *compressus* are not those so named by earlier writers.

Sir H. T. de la Beche, among other proofs of his attention to the organic contents of the Lias of Dorsetshire, presented to the Geological Society a drawing of a rare Belemnitic fossil, which he termed an Orthoceratite (1829). It has but a small and slender guard, and a very slender phragmocone, but possesses an elongated anterior shell. Prof. Huxley has described it from more perfect examples, under the generic title of Xiphoteuthis. It occurs in the middle Lias of Lyme Regis.[1]

In the first volume of my work entitled 'Illustrations of the Geology of Yorkshire,' 1829, I noticed nine specific groups of Belemnites from the Lias and Oolites of the Yorkshire coast, from my own observations among the cliffs, and still more from the rich collections freely opened to be by Mr. Bean and Mr. Williamson. The public collections at Whitby and York were also examined. Almost immediately after the publication of this volume I visited the Strasburg Museum, and found M. Voltz busy in those excellent observations which place him, in my judgment, in the very first rank of the naturalists who have really studied Belemnites. From him I learned much; and in the second edition of my work already named (in 1835) the following stands as the "Synoptic List of Yorkshire Species," arranged according to the strata in downward series:

CHALK	B. mucronatus, *Sow.*
„	— granulatus, *Sow.*

---

[1] 'Mem. Geol. Survey' ("Organic Remains"), 1864.

Red Chalk	B. Listeri, *Phil.*.
Speeton Clay	— minimus, *Sow.*
„	— jaculum, *Phil.*
Speeton Clay and Kimmeridge Clay .	— lateralis, *Phil.*
Coralline Oolite	— abbreviatus, *Mill.*
„	— Blainvillii, *Voltz.*
Oxford Clay	— gracilis, *Phil.*
Kelloways Rock	— anomalus, *Phil.*
„	— tornatilis, *Phil.*
Bath Oolite	— quinquesulcatus, *Bl.*
„	— Aalensis, *Voltz.*
Upper Lias	— tubularis, *V.* and *B.*
„	— compressus, *Voltz* not *Sow.*
„	— trifidus, *Voltz.*
„	— subaduncatus, *Voltz.*
„	— paxillosus, *Voltz.*

Mr. Samuel Woodward recorded in his 'Synoptical Table of British Organic Remains' (1830) the following species:

Chalk	B. granulatus, *Sow.*
„	— electrinus, *Mill.*
Chalk Marl.	— lanceolatus, *Sow.*
Gault	— Listeri, *Mant.*
„	— attenuatus, *Sow.*
„	— minimus, *Sow.*
Oxford Clay	— gracilis, *Phil.*
Stonesfield Slate	— fusiformis, *Park.*
Great Oolite	— compressus, *Sow.*
Inferior Oolite	— sulcatus, *Mill.*
„	— ellipticus, *Mill.*
„	— abbreviatus, *Mill.*
Lias	— tubularis, *V.* and *B.*
„	— elongatus, *Mill.*
„	— aduncatus, *Mill.*
„	— longissimus, *Mill.*
„	— pistilliformis, *Sow.*
„	— acutus, *Mill.*
„	— penicillatus, *Sow.*

Dr. Buckland's researches in Belemnites began in 1829.[1] In his 'Bridgewater Treatise on Geology and Mineralogy,' 1836, he gave many interesting drawings of Belemnites and other fossil Cephalopoda, and added useful comparative diagrams of recent decapod and octopod cuttles. He gives many examples of "ink-bags of Belemno-sepia, and their

[1] 'Geol. Soc. Proc.,' i, 96; and 'Trans.,' iii, 217.

nacreous sheaths," from the Lias of Lyme Regis, and represents one specimen, stated to be unique, from the cabinet of Miss Philpotts, which shows together the guard, phragmocone, and ink-bag; he names it *B. ovalis.* Of *B. pistilliformis,* Sow., also, a specimen is figured, showing traces of the ink-bag. A third Belemnite from the Lias, very short, with thin guard, is called *B. brevis*? A restoration of " Belemno-sepia," with the included shelly parts, usually called Belemnite, is attempted, and the analogies of the Belemnitic animal are discussed in the spirit of the remarks of Blainville.

In consequence of the cutting through Oxford Clay at Christian Malford, in Wilts, by the Great Western Railway, Mr. W. C. Pearce, Mr. S. P. Pratt, Mr. Buy, Mr. Reginald Mantell, Mr. W. Cunnington, and other collectors, were able to obtain many admirable examples of two groups of Cephalopoda, which gave occasion to several valuable memoirs. In January, 1842, Mr. Pearce noticed a conical shell which resembled the Belemnite by having, like it, a concamerated portion traversed by a marginal siphuncle, and protected by a brown, thick, shelly covering, which gradually became thinner towards the superior part. " Immediately above the chambers is an ink-bag, resting on what resembles the upper part of a sepiostaire, and composed of a yellow substance finely striated transversely, being formed of laminæ of unequal density." Some other notices are added, and the author assigns to his specimens the generic name of Belemnoteuthis.[1] Mr. Cunnington afterwards added the specific name. It has been regarded as congeneric with *Acanthoteuthis* of Münster, and called *A. antiquus.*

The 'Philosophical Transactions' for 1844 contain the important memoir of Prof. Owen on the Belemnites found in the Christian Malford cutting. Many parts of the structure of Belemnites, and the affinity of the animal to other Cephalopoda, are successfully cleared up in this instructive and valuable essay. The author was, however, not supplied at that time with such satisfactory specimens of true Belemnites as of the Acanthoteuthidæ already mentioned. In consequence, these groups were not distinguished, and the beautiful restored figure ('Phil. Trans.,' 1844, pl. viii) is not so satisfactory as later discoveries might have furnished the means of producing. To this memoir, and to a later work of the professor,[2] we shall refer in future pages for information on special points of structure.

In 1848, Dr. Mantell communicated to the Royal Society "Observations on some Belemnites and other Fossil Remains of Cephalopoda, discovered by his son, Mr. Reginald Mantell, C.E., in the Oxford Clay at Christian Malford, near Trowbridge, in Wiltshire." Two species of straight-shelled Cephalopoda are here figured and described, and shown to belong to two genera, viz., one the Belemnoteuthis of Pearce, with ink-bag, uncinated arms, and a chambered shell, with thin investing fibrous sheath; the other a Belemnite common in the Oxford Clay, to which several names have been assigned. Now, for the first time, the two narrow processes which extend beyond the divisions of the phragmocone

---

[1] 'Proceedings of Geol. Soc.,' vol. iv, 592.

[2] 'Palæontology,' 8vo, ed. 2, 1861.

were clearly distinguished, and the several parts of this somewhat complicated shell placed in comparison with the corresponding parts of Belemnoteuthis. In 1849, further details from other examples were communicated by the same indefatigable naturalist to the Geological Society, showing the two remarkable processes already alluded to. The author expressly declares that he had never found in specimens of the numerous Belemnites which he had examined any trace of the ink-bag or its dark contents.

The volume for 1850 of the 'Memoirs of the Palæontographical Society' contains an account of the Cephalopoda of the Stonesfield Slate by Prof. Morris and Mr. Lycett. Two species are described and figured, viz., *B. fusiformis*, Park., and *B. Bessinus* D'Orb., which is regarded by Oppel as a synonym of *B. sulcatus*, Miller, and *B. canaliculatus*, Schl.

In the volume for 1853 of the same Society the Belemnitidæ of the Upper Cretaceous system are described and figured by Mr. Sharpe, five species, which will be found with the same names in the Catalogue of Prof. Morris.

Prof. Morris presents in his 'Catalogue of British Fossils,' edition 2nd, 1854, the following thirty-one species :

CHALK	Belemnitella lanceolata, *Schl.* = *Belemnites mucronatus*, Sow.
,, ...	— mucronata, *Schl.* = *B. electrinus*, Mill.
,, ...	— plena, *Blain.* = *Actinocamax vera* of Miller, and *Belemnites lanceolatus*, Sow.
,, ...	— quadrata, *Defr.* = *B. granulatus*, Sow.
GAULT AND RED CHALK .	Belemnites ultimus, *D'Orb.* (*B. Listeri*, Phil.). Folkstone.
GAULT	— attenuatus, *Sow.* Folkstone.
,, ...	— minimus, *Lister.* Ringmer ; Bletchingly ; Folkstone ; Wilts.
SPEETON CLAY . . .	— lateralis, *Phil.* Brantingham ; Speeton.
,, ,, . .	— jaculum, *Phil.* Speeton.
CORALLINE OOLITE . .	— abbreviatus, *Mill.* Malton ; Garsington.
OXFORD CLAY AND KELLOWAY ROCK	— anomalus, *Phil.* Yorkshire.
,, ,, ,, .	— Beaumontianus, *D'Orb.* Loch Staffin.
,, ,, ,, .	— gracilis, *Phil.* Yorkshire.
,, ,, ,, .	— hastatus, *Bl.* Cambridgeshire ; Wiltshire ; Dorset.
,, ,, ,, .	— Owenii, *Pratt* (*B. Puzosianus*, D'Orb.). Wilts ; Dorset.
,, ,, ,, .	— tornatilis, *Phil.* Yorkshire.
LOWER PART OF THE GREAT OOLITE	— Bessinus, *D'Orb.* Stonesfield.
,, ,, ,, .	— fusiformis, *Park.* (*B. Fleurisianus*, D'Orb.). Stonesfield.
INFERIOR OOLITE . .	— canaliculatus, *Schl.* Gloucestershire.
,, ,, . .	— ellipticus, *Mill.* Dundry.
,, ,, . .	— giganteus, *Schl.* (Morris gives for synonyms *B. quinquesulcatus* of Blainville, *B. compressus* of Sow., and *B. Aalensis* of Voltz. In some foreign museums *B. longus* of Voltz is added to this synonymy, which will be considered hereafter.)

INFERIOR OOLITE . . .	B. sulcatus, *Mill.*	Considered to be the equivalent of *B. apici-conus*, of Blainville. Dundry.
LIAS	— acuarius, *Schl.*	(It is supposed by Prof. Morris to be identical with *B. tubularis* of Young and Bird and *B. lagenæformis* of Zieten.) Whitby, Yorkshire ; Gloucestershire.
„	— acutus, *Mill.*	Shornecliff ; Charmouth.
„	— aduncatus, *Mill.*	Lyme ; Weston.
„	— breviformis *Voltz.*	Gloucestershire.
„	— brevirostris, *D'Orb.*	Cheltenham.
„	— compressus, *Voltz.*	Yorkshire.
„	— elongatus, *Mill.*	Somerset ; Dorset ; Ross ; Cromarty.
„	— longissimus, *Mill.*	Lyme ; Western ; Boll.
„	— paxillosus, *Voltz.*	Yorkshire ; Gloucestershire.
„	— penicillatus, *Bl.*	Dorset ; Gloucestershire.
„	— pistilliformis, *Sow.*	Charmouth.
„	— subaduncatus, *Voltz.*	Yorkshire.
„	— trifidus, *Voltz.*	Yorkshire.

In 1855 Mr. S. P. Woodward, besides giving a compendious and effective classification of the Belemnitidæ as a group of Cephalopoda Tetrabranchiata, presented a drawing of the solid parts of the Belemnitic animal seen dorsally,[1] and another of *Belemnoteuthis antiquus* seen ventrally.[2] In his supplement (1856) he calls attention to a specimen in the possession of Mr. Buckman which exhibits the fossil ink-bag within the phragmocone of a Belemnite. The classification here referred to will be considered hereafter.

The work of Mr. Martin Simpson (1855) on the 'Fossils of the Lias of Yorkshire' includes notices of forty species and varieties of Belemnites, of which specimens may be seen in the Whitby Museum. I have examined that collection carefully, and have found it very instructive. Mr. Simpson's descriptions are careful, but, being unaccompanied by figures, inspection of the original specimens is necessary for understanding the distinctions relied on.

For the latest and most important additions to our knowledge of the structure of Belemnitidæ by British authors we are indebted to Prof. Huxley's descriptions of some remarkable specimens obtained by Mr. E. H. Day from the Lower Lias near Charmouth. In this elaborate memoir[3] the facts embodied in the various statements of previous observers are re-examined, and a nearly complete view is furnished of the structure and relations of the solid parts of the Belemnitic animal. In particular, the anterior solid parts (*pro-ostracum*) are placed in their true significance, both in the Belemnites of ordinary form, as *B. elongatus*, and in the new genus Xiphoteuthis. To this valuable essay we shall again refer.

[1] 'Manual of the Mollusca,' pl. ii, fig. 5.

[2] Ibid., p. 75.

[3] "British Organic Remains" ('Memoirs of the Geological Survey'), Monograph 2, 1864.

Several foreign writers of eminence have incidentally noticed the more frequently occurring British Belemnites, and in this manner, better sometimes than by the descriptions or figures, we are able to establish, in a few instances, a satisfactory synonymy with Schlotheim, Blainville, Voltz, D'Orbigny, Münster, Quenstedt, Oppel, and Von Meyer. To these authors reference will be frequently made as we proceed.

Lists of Belemnites, and scattered notices of individual species, have been often given by British authors while describing particular strata or remarkable localities; but there is, perhaps, no group of fossils which demands so much caution in quoting references of this nature to species not always really determined by examination of a large number of specimens in different periods of growth.

STRUCTURE OF BELEMNITIDÆ.

According to the knowledge now acquired, the solid parts of the Belemnitic animal consisted of an internal shell of somewhat complicated structure, which for clear explanation requires representations of at least three aspects of the surface and three or more sections of the interior. By employing always fixed letters of reference for these aspects and sections, the descriptions become more compact, symmetrical, and characteristic. To begin with the principal sections, which are represented in the first diagram, one being longitudinal, the others transverse—

The portion of the Belemnite most commonly preserved in a fossil state is the posterior or caudal part, marked G in the first diagram. It is to this sparry, usually pointed, and more or less cylindrical, conical, or tapering mass, that the ancient discussions as to the horny, shelly, or mineral nature of the Belemnite belong. This was, in fact, for Woodward and Lister the Belemnite. It has been called the guard, sheath, rostrum, or osselet.

In the transverse section taken at the apex of the phragmocone the concentric curves represent successive stages of growth, the inner rings being deposited first, as in exogenous trees. Indications of a thin capsule, or formative membrane, appear in some Belemnites of the Oxford Clay and Lias, investing the guard.[1] In the Oxford Clay it is represented by a granular incrustation; in some Lias Belemnites it appears in delicate small plaits, like ridges and furrows. The whole mass of the guard is fibrous, and usually transparent carbonate of lime; not exactly calcareous spar, unless we suppose a minutely fibrous kind of crystallization, like that sometimes seen in stalactite. It is still more like arragonite, and some kinds of recent shells, not such as Pinna or Hinnites, with large prismatic cells, but, as Dr. Carpenter remarks, like Septaria. A distinctly cellular

---

[1] Mantell, 'Phil. Trans.' Owen, 'Phil. Trans.' Huxley, 'Mem. of Geol. Survey.'

structure is difficult to demonstrate in these fibres, which Owen describes, in specimens from Christian Malford, as of a trihedral prismatic form, and of $\frac{1}{2000}$th of an inch in

DIAGRAM 1.

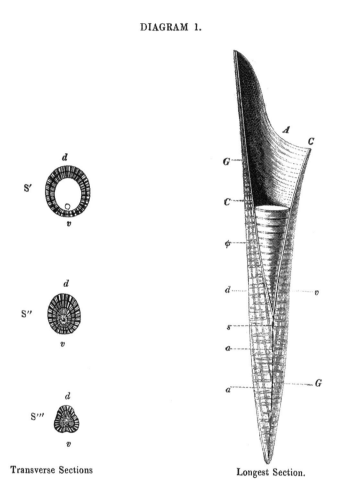

Transverse Sections   Longest Section.

G. Guard. φ. Phragmocone. C. Conotheca. A. Alveolar cavity. s. Sphericle of the guard. a a. Axis of guard.
S′ S″ S‴. Transverse sections. d. Dorsal aspect. v. Ventral aspect.

diameter. In some species they are very much larger. They appear sometimes connected into pencils; mostly, however, they are ranged perpendicularly to the laminæ of growth, which they cross without interruption, to meet on the axial, or, as it is called by Voltz, the apicial, line. The Belemnite is found in nearly the same mineral condition whatever be the containing rock. It is sometimes wholly removed, leaving a hollow, into which the conical mould of the alveolar cavity projects. This happens in the upper beds of the Lias in Yorkshire and in the Marlstone of Rutland.

The Belemnitic shell begins in a very small spherule, the wall of which has not been very clearly seen; but, by observation of specimens in Oxford Clay and Lias, it appears to have been a thin cellular tissue (*s'*) like that of the youngest parts of the conotheca and the smallest septa.

DIAGRAM 2.

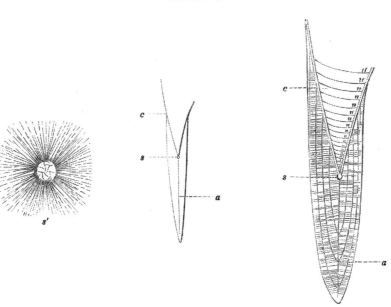

It is supposed by some writers to have been extended backward into a small canal at the meeting of the fibres along the axis of the guard, and forward into the pipe or siphuncle of the phragmocone, of which more hereafter. On the little spherule a system of pencils of radiating cells grows up, in a lamina which embraces the spherule on all sides except its anterior or part of its anterior face. There the phragmocone is already begun, either by the spherule extending into a short tube representing the siphuncular opening in a septum, or without such entering short tube. Successive laminæ of radiating cells, more and more extended backward, but stopped in a forward direction by the growing phragmocone, cover the spherule, meet on the axial line, and constitute a small, rather fusiform Belemnite (middle figure), in which the septa appear, and the siphuncle takes its place.

In some instances there appear to be traces of more than one such spherule, with its pencils of diverging fibres, enclosed in the guard, as if the phragmocone had aborted, or its earliest chambers were spherical, and united by a short pipe.

If this description be well considered, it will appear certain that the phragmocone is not a chambered body made to fit into a conical hollow previously formed in the Belemnitic sheath, as some have conjectured, but that both sheath and cone grew together; or, if any difference there were, the phragmocone must have been of earlier growth, and by its enlargement limited the forward extension of the successively deposited laminæ of the guard.

The phragmocone was formed on the exterior of a secretive surface, and the fibrous sheath on the interior of another secretive surface, both extended together from the germinal capsule. The incremental striæ are external on the phragmocone, and internal on the guard.

The guard is sometimes fusiform when young, and grows cylindrical or conical with age; but it is doubtful if any example is satisfactorily in accordance with the figure of Blainville (pl. i, fig. 4), which represents a fusiform mass enclosed in laminæ of growth quite parallel to it. In some cases detached Belemnites assume a fusiform shape by the decay of the laminæ in the peripheral parts about the point of the alveolus. This seems to arise from a difference of composition of the laminæ thereabout. It is seen very often in the elongated Belemnite of the Speeton Clay, and in specimens of *Belemnites quadratus*. On such as these Miller commonly founded his genus Actinocamax.

DIAGRAM 3.

W

The part marked W is often wasted away in soft, brittle laminæ.

The surface of the guard is sometimes marked by ramified impressions of vessels, and more frequently by certain systematic grooves, which furnish some of the grounds for convenient classification.

Confining our attention at present to the guard, it may be remarked that the anterior edge is seldom completely traceable; we may, however, be sure that in every case the dorsal edge and its alveolar cavity extended much farther forward than the ventral edge, as represented in Diag. 1. Another point of importance regards the plane of symmetry of the shell, for this always passes directly through the axis or apicial line of the guard, from the middle of the dorsal to the middle of the ventral face. On each side of this plane (except in cases of deformity or accident) the masses and areas are equal and equally disposed, and the markings of the surface are in pairs, while along the middle line thus defined on the dorsal and ventral faces the markings are single. Thus, three aspects of the guard always require attentive inspection and special description, a circumstance rarely observed by Belemnitologists.

The generally smooth surface of Belemnites is only broken by grooves of greater or less depth, striæ, and small plications, and continuous or interrupted small ridges. All these correspond to peculiarities of the formative membranes, and are sufficiently characteristic and constant to serve for more than specific distinction, as will more fully appear hereafter.

The successive laminæ of the guard are not parallel, but in the greater number of cases grow thicker towards the point, and remarkably thinner towards the anterior edges. These laminæ meet the apicial line (*a*) at angles usually acute, and still more acute in their intersection with the conical cavity, which receives the phragmocone, so that they might even appear to be parallel to it. As many as three hundred laminæ of growth have been counted in the cross section of the solid part of a Belemnite from the Oxford Clay, and as many as eighty in the section of the thinner guard over the alveolar cavity.[1] The thickness of the laminæ is about $\frac{1}{500}$th of an inch on the average, sometimes not more than $\frac{1}{1000}$th of an inch.

DIAGRAM 4.

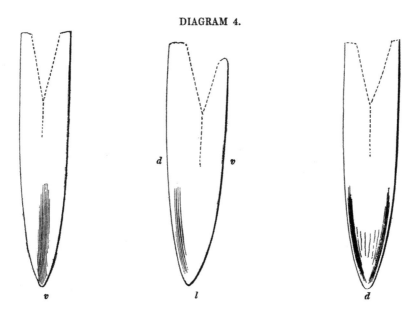

v Is the ventral or lower aspect, *l*, the lateral, and *d* the dorsal, of *Belemn. canaliculatus*, Oxford Clay.

In some Belemnites in the calcareous nodules of the Lias, and in others which occur in the hard Chalk of Antrim, it is difficult to see the laminæ of growth in the clear, horny, yellow, spathose mass of the guard. On the other hand, specimens which occur in clay, shale, or open-grained limestone, show these laminæ very clearly and in great plenty.

DIAGRAM 5.

Longitudinal section of
laminæ of guard.

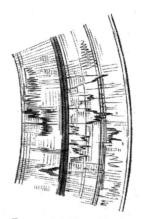

Transverse section of laminæ
of guard.

[1] Owen, 'Phil. Trans.,' 1844.

PHRAGMOCONE.

The conical cavity of the guard is often empty, but as often filled with a shell of similar form, expanding gradually forward, and divided across by many shelly plates, each concave outwardly, and pierced by a small pipe or siphuncle (σ) near one edge. By these plates or septa the conical shell is divided into chambers, the last being very large in comparison with the others, and destined to cover the breathing organs, heart, and other viscera. The thin conical shell (*c* in the diagram) which covers these chambers is distinct from the substance of the guard, and is called the 'conotheca' by Huxley. The whole chambered mass with this shell is named "*phragmocone*" by Owen. It has been also called "*alveolus*," but that title is better bestowed on the cavity in the guard which receives the phragmocone. The sheath or guard extends forward over this conotheca, but grows by degrees so thin as to become untraceable. Beyond the end where it thus disappears the conotheca is further extended, and in some instances acquires so much length and peculiarity of form as to require a separate designation. The most convenient, perhaps, is that proposed by Huxley, viz., "*pro-ostracum*," or anterior shell. In some cases this extension seems to run out in one simple broad lobe—this appears to happen in Lias Belemnites; in others—Oxford Clay species, for example—it forms two long, narrow, parallel plates. Whatever be the form of this *pro-ostracum*, it is properly a dorsal extension of the conotheca of the Belemnite. In the genus *Xiphoteuthis* it is a very elongated part, larger than the guard, and united to it by a very long (generally depressed), shelly, conothecal extension.

The phragmocone is rarely found complete. The best examples to show its general

DIAGRAM 6.

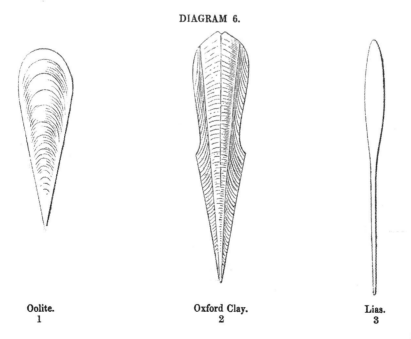

Oolite.
1

Oxford Clay.
2

Lias.
3

figure have been obtained from the laminated Oolite of Solenhofen, the laminated Clay of Christian Malford, and the laminated beds of Lias in Dorsetshire and Yorkshire. Three different forms are thus put in evidence ; we may add, as a fourth, the elongate shape of Xiphoteuthis.

In the preceding sketches the conotheca is supposed to be opaque, so that the interior septa are not seen. The aspect is dorsal. 1 is from *Belemnites hastatus*, from Solenhofen. 2, *B. Owenii*, from Christian Malford. 3, *Xiphoteuthis elongatus*, from Lyme.

The external aspect of the phragmocone of *B. paxillosus* is given in the next diagram, in three aspects, dorsal, lateral, and ventral. The figures are *restored*, from a variety of specimens which give the requisite information, no one being complete.

DIAGRAM 7.

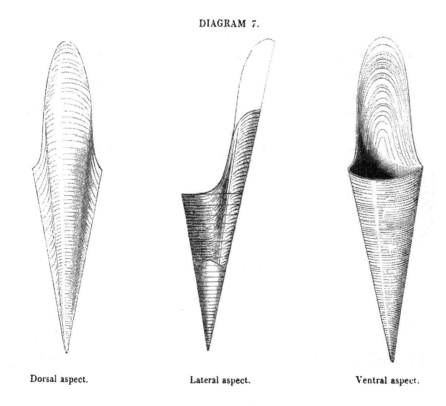

Dorsal aspect.   Lateral aspect.   Ventral aspect.

The surface of the conotheca is marked by lines of growth in singular and characteristic curves, first accurately examined by Voltz.[1] According to this observer, the conical surface may be described in four principal regions radiating from the apex ; one dorsal, with loop lines of growth advancing forward ; two lateral, separated from the dorsal by a continuous straight or nearly straight line, and covered with very obliquely arched lines in a hyperbolic form, in part nearly parallel to the dorso-lateral boundary-line, and in part reflexed, so as to form lines in retiring curves across the ventral portion, nearly

[1] 'Observations sur les Belemnites,' 1830.

parallel to the edges of the septa. Measured round the cone, the dorsal region occupies about ¼th, each lateral (or hyperbolic region, as Voltz calls it) about ⅛th, and the ventral region about ½ the circumference. The straight or nearly straight boundaries of the dorsal space are called asymptotes. There is occasionally a straight line along the middle of the dorsal region crossing the "ogivial" dorsal axis; this line is in some cases elevated, in others depressed. Fine longitudinal striæ appear on the outermost layer of the phragmocone especially in the ventral region.

The proportions of the phragmocone, its transverse septa, chambers, and siphuncle, vary in different species of Belemnites. The axis of the general figure is for the most part a little curved, always in such a manner that the apex is slightly bowed towards the lower or ventral side. If a longitudinal section be made right through the siphuncle, from the dorsal to the ventral line, and the angle of inclination to each other of the sides of the phragmocone be measured, it will be found that this angle grows more and more open as we approach the apex. If another longitudinal section be taken across the cone from side to side the boundaries will be nearly straight, and the angle of inclination of the sides nearly constant. It is this angle which should be given as a part of the specific character. Taking an example from a phragmocone of any Belemnite, the septa are found to be nearer and nearer to one another as the cone grows smaller and we approach the apex. In some Belemnites the septa are circular; in others elliptical, the longer axis being between the dorsal and the ventral face, and passing through the siphuncle. In *B. paxillosus* the diameters are as 100 to 108. The curvature of the septal plate is nearly spherical or spheroidal; the arc included is about 50°, but toward the apex greater, even exceeding 90°. The radius grows shorter and shorter in a greater proportion than the diameter lessens. The intervals between the septa are about ⅙th of the diameter. In a series of septa exposed by section of the phragmocone of *B. canaliculatus*, the intervals were thus found nearly—

17, 16, 15, 14, 13, 12, 11, 10, 10, 9, 8;

and in another less regular—

15, 14, 13, 10, 8, 10, 9, 9, 8.

In neither case could the septa be counted to the end, but the whole number was estimated to be about thirty in a length of ¾ths of an inch. In a specimen from Lyme Regis ten septa occur in the space of 1/10th of an inch near the apex.

M. Voltz figures a fragment of *Belemnites Aalensis* containing ten septa in a space of 1·83 inch. The angle of this cone being 21°, and its base 1·14 inch, it is practicable to calculate the probable number of septa, till near the apex they would become untrace able. The cone would be 3¾ inches long, and would contain fifty septa before the smallest was reduced to 1/10th of an inch, and sixty before the diameter was reduced to 1/18th.

In a pyritous specimen of *B. vulgaris*, Y. and B., from the Upper Lias of Whitby I count sixty septa, of which the anterior twenty are singularly pressed inwards close up to the undisturbed part, which is only ¾ths of an inch long.   In the Oxford Museum is a portion of the phragmocone of *B. giganteus*, which has a length of 3⅓ inches, greater diameter 2¹⁄₂, lesser 1⅝, and ten septa.   If continued to the apex, its length would have been above 8 inches, the number of septa eighty or more, before becoming untraceable.   The anterior vaulted part of the phragmocone may be estimated as equal to the chambered part ; and adding to this the probable extent of the guard, the total length of the shell would be not less than 24 inches.

This, however, is probably not the utmost length of Belemnites, such as those called *giganteus*, *Aalensis*, and *ellipticus*, in which sometimes the sparry guard behind the phragmocone attains the length of a foot or more, and the total extent may have been 3 to 4 feet.

Fifty septa occur in a specimen of *B. paxillosus*, in the Bristol collection, in a length of 2·8 inches.

The angle of inclination of the sides of the phragmocone is probably constant for the same species of Belemnites, but it varies much in different specimens.   Quenstedt gives a section from Lyme Regis of a phragmocone with sides inclined at 8° 40′.[1]   I have never met with so slightly converging a cone in the Lias ; but among Mr. Moore's specimens from Ilminster, usually referred to *B. elongatus*, is a cone with slopes meeting at 12°. It is more common to find the angle above 20° and under 28° for all species in the Lias and Oolites.   I have a specimen measuring 32° ; unfortunately we cannot often refer these phragmocones to their proper guards.

The structure of the conotheca and of the transverse septa has not appeared the same to every observer, partly because there may be some real differences in the shells of different species, but more frequently from incomplete knowledge of the disguises induced by laminæ of bisulphide of iron and carbonate of lime attached to the real shell-structure. To M. Voltz must again be awarded the praise of just observation in this matter.   In his 'Treatise on Belemnites,' an instructive figure is given, to show the difference between the description of Miller and his own.   This is copied in diag. 8, where M shows the section according to Miller, V that according to Voltz.

DIAGRAM 8.

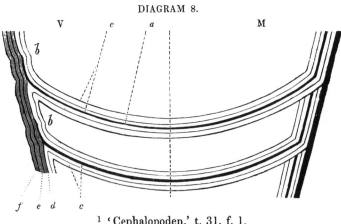

[1] 'Cephalopoden,' t. 31, f. 1.

Miller supposed the transverse plate of the phragmocone to be composed of five parallel laminæ, all bent forward to line the conotheca, which consisted of one thin plate only (M, in diag. 8). Voltz explains that the conotheca is formed of three laminæ; that it is bordered internally by only a short flange extended from the transverse plate; that it is inflected at the junction with this flange, the inflection being limited by two small incisions. The transverse plate (a, in diag. 8) is composed of several thin laminæ, but is not so thick as the conotheca. Miller has not distinguished from the real substance of the shell the thin laminæ of carbonate of lime (diag. 8, c) which often line internally the conotheca and septa. My own observations agree in general with those of M. Voltz, but it seems useful to describe the appearances presented in two or three easily procurable species, which present some small differences.

Diag. 9 represents the section along the dorsal face of three transverse septa abutting on the conotheca. This enclosing shell is formed of three layers towards the opening, but of two only towards the apex. Still nearer to the apex it is composed of one layer only, and that finally ends in a porous or cellular plate, very like what can be traced on the limiting texture of the spherule at or beyond the apex of the phragmocone. This Belemnite is in the Oxford Museum, from the Lias of Lyme Regis. The septa are flanged, and, as it were, bracketed where they meet the conotheca, and a small triangular interstice appears between the bracket and the conotheca.

In *Belemnites vulgaris* of Young and Bird, from the Upper Lias of Whitby, the septa are apparently single plates—at least this appears to be the case in the hinder part of the cone.

DIAGRAM 9.

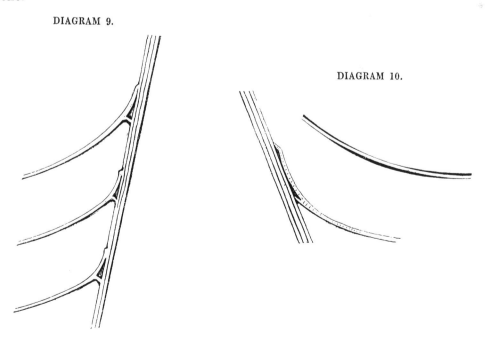

DIAGRAM 10.

In diag. 10, taken from another specimen in the Oxford Museum, from Lyme

Regis, the long advancing flange is clearly traced against the triple conotheca. The septum itself is of unusual construction, as represented in the second figure. It contains three plates—the middle one clear and sparry, and of uniform thickness; the upper one is quite dark, and thicker in the middle, like a periscopic convex lens; the lower one is thinner in the middle, like a corresponding concave lens.

DIAGRAM 11.

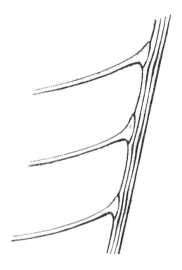

In a specimen of *B. tornatilis* from the Oxford Clay the conotheca is triple, the two inner laminæ being somewhat curved in conformity to the short abutment of the septa (diag. 11). The appearances which have been described are not always easy to eliminate from the various shaded and crystallized laminæ which overlie the septum, and present delusive appearances of structure. Bisulphides of iron, zinc, and lead, and carbonate of lime in a variety of aspects, form layers on the walls or fill up entirely the chambers.

According to my observations, there is reason to expect that the phragmocone will afford specific characters more definite, if not so often available as those of the guard.

When the conotheca is removed, and the cast of the chambers appears, the impressions of the septal flanges remain in several species very plainly on the cast, and cause undulations in the exterior outline of the conotheca.

Through each transverse plate is a perforation, near the ventral margin, formed by the retroflection there of the laminæ of the plate. These reflected parts of the plate are some-

DIAGRAM 12.

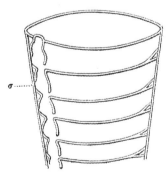

DIAGRAM 13.

times found to be expanded in the interseptal spaces, as happens to many Orthocerata and Nautili. This is represented by Voltz in a specimen of *B. Aalensis*.[1] The series of

[1] 'Observ. sur les Belemnites,' pl. i, fig. 3.

perforations with the short tubes to each constitutes what is called the siphuncle (σ), which is sometimes so close to the ventral side that its expansions touch the conotheca.[1] In diag. 12 is seen the longitudinal section through the siphuncle, showing the retro-flexions of the septa which form the siphuncle, and how this approaches and touches the conotheca. The phragmocone is that of *Belemnites vulgaris,* from the Upper Lias of Yorkshire. Diag. 13 shows the marks left by the siphuncle on the "chambered cone," which remains when the conotheca is wholly or partially removed.

As already observed, the specimens are few in which the phragmocone and guard are found together complete, or in such a state as to allow of a correct judgment of the whole figure, if complete. Such specimens have been found in the Oolite of Solenhofen and in the Oxford Clay near Chippenham. In the Lias of Lyme Regis and Yorkshire the structure of the phragmocone in relation to the guard is sufficiently ascertained to justify a restoration of the whole shell. In diag. 14 are presented three such figures, one from the

DIAGRAM 14.

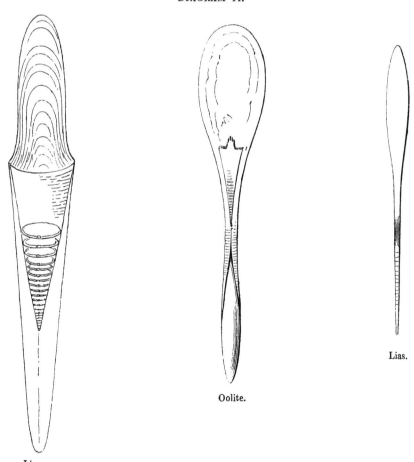

Lias.

Oolite.

Lias.

Oolite of Solenhofen (*B. hastatus*), two from the Lias, viz. *B. paxillosus* and *Xiphoteuthis elongata.*

[1] Voltz., loc. cit., pl. i, fig. 1.

Though no complete specimen has occurred of any Belemnitic animal, late discoveries, especially those of Mr. Day in the Lias at Lyme Regis, have established many facts regarding the muscular system of the body, the hooks on the arms, the funnel, ink-bags, eyes, and other parts of importance. Restorations have been attempted at various times by several naturalists since first Miller presented a general sketch showing the affinity of the Belemnite to the Sepiaceæ. Buckland's figure ('Bridgewater Treatise,' pl. lxi, fig. 1) represents the ventral aspect of the animal, with funnel, ink-bag, and posterior latero-caudal fins. Only eight arms are distinct in the figure. On the two longer ones are circular suckers. Quenstedt ('Cephalopoden,' t. xxiii, fig. 16) presents a drawing of the dorsal aspect, showing lateral fins expanded from the alveolar region, with a thin membranous expansion over the fibrous guard, and an equally thin fleshy covering to the anterior part of the phragmocone. The two longer arms are bare, the eight shorter ones have hooks. D'Orbigny sketches a side view, with pointed fins near the extremity of the tail, the Belemnite lying exposed on the back, the two longer arms bare, the others with two rows of suckers each. Ten arms are assigned to the animal by Quenstedt and D'Orbigny, as in the restorations by Owen,[1] which, however, is partly modelled on Belemnoteuthis. This uncertainty in regard to the prehensile organs of the animal has been in some degree removed by the researches of Mr. Day in the Belemnitic beds of Lyme Regis and Charmouth. Several specimens collected by this gentleman have shown in their true relative position the guard, the phragmocone, the anterior extension of the conotheca, and the coronet of hooks which margined the arms. Two of these specimens have been

DIAGRAM 15.

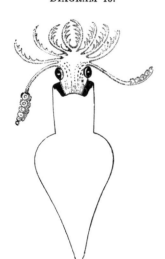

figured by Huxley.[2] In one the part of the guard below the phragmocone measures about $\frac{1}{10}$th part of the whole animal, in the other about $\frac{1}{7}$th. Von Buch had estimated the length of the animal to exceed that of the shell by eight or ten times; but if by the term shell we mean to express the whole of the solid calcareous or horny substance, its length is nearly equal to that of the animal, for it reaches, as in the Sepiaceæ, to the edge of the funnel. The outline of the body of the animal is not yet recovered; in specimens of the Lias the ink-bag, funnel, and portions of the sclerotic arcs of the eye, are designated by Huxley; but the muscular substance of the arms, mantle, and fins is untraceable. Such traces do occur in the Oxford Clay in connection with Belemnoteuthis, but the entirely different proportions of the guard in these animals render it improbable that the swimming apparatus could be quite on the same model as in the Belemnite.

---

[1] Owen, 'Phil. Trans.,' 1844.   [2] 'Mem. of Geol. Survey,' 1864.

Especially in the anterior part and in the circle of hooked arms Mr. Day's specimens have been found very suggestive. The arms. are, on the whole, unexpectedly short (as compared with those of Belemnoteuthis) ; the head also, with the eyes, appears smaller in proportion than among other Cephalopoda, and the whole body seems longer in proportion than is usual among them, in this respect more resembling the Calamaries (Loligo) than the Cuttles (Sepia).

CLASSIFICATION OF THE BELEMNITIDÆ.

To the earlier writers the family of the Belemnitidæ was known only by the two prominent examples now called Belemnites and Belemnitella, and of these they knew not the whole. The chambered cone was not always clearly understood as an essential and very characteristic part of the whole shell until Klein (1731) gave forth his 'Descriptiones Tubulorum Marinorum,' which include Orthoceratites. In this essay Belemnites constitute the eleventh genus of Tubuli Marini, and are thus clearly defined:

" Belemnites est tubulus marinus ; fossilis ; materiæ ad Seleniticam accedentis ; teres ; transversim fractus concentricis striis, in longitudinem fissus canaliculo pervio, semper in medio posito, donatus ; in basi nonnunquam ferens conum, olim testaceum, concameratum, instructum siphunculo."

This cone was, and still often is, called the alveolus, though that name properly belongs to the conical cavity. The siphuncle of this chambered shell is regarded by Klein as connected with the central canal of the guard, and with a globule at its apex. He founds this opinion chiefly on specimens from the Chalk, and compares this structure to that of the chambered *Nautilus crassus*, distinguishing this from the open-shelled Argonaut. In agreement with previous writers, he distributes Belemnites into three groups—cylindric, conical, fusiform.

Fifty years afterwards, Miller, satisfied himself of the affinity of Belemnites to the shelly parts of Cephalopoda, perceived the analogy of the animal to which the fossil belonged with the recent possessor of the " Cuttle-fish bone," and presented a conjectural drawing to illustrate the generic character, which is in these words:

Genus. Belemnites.—A cephalopodous? molluscous animal, provided with a fibrous, spathose, conical shell, divided by transverse concave septa into separate cells or chambers connected by a siphuncle, and inserted into a laminar, solid, fibrous, spathose, subconical or fusiform body extending beyond it, and forming a protecting sheath or guard. (' Geol. Trans.,' 2nd series, vol. ii, p. 48.)

It is unnecessary now to notice his genus " Actinocamax," which is only the retral portion of the guard of a Chalk or Gault Belemnite, separated at or near the apex of the phragmocone—often by natural decay of the shelly laminæ.

De Blainville's great work follows that of Miller, and his classification is in all respects more advanced and comprehensive. It contains eight sections, of which the first is founded on a mistake of Miller, and the last is merely a small appendix of forms which are now distributed in the other sections.

A. No alveolar cavity. (*Actinocamax* of Miller.)

B. Alveolar cavity very small, fissured on the margin, and without septa. (Example —*Belemnites quadratus*. Chalk.)

C. Alveolar cavity large, fissured on the margin, and without septa. (Example— *Belemnites mucronatus*. Chalk.)

D. Alveolar cavity large, chambered, siphunculated, with a ventral canal more or less evident from the base to the summit of the guard. (Example—*B. sulcatus*, Miller. Oolite.)

E. Alveolar cavity large, chambered, siphunculated ; without fissure or canal at the base, but with two lateral furrows at the summit of the guard. (Example—*B. apici- curvatus*, Blainv. Lias.)

F. Alveolar cavity large, chambered, siphunculated ; no fissure or canal at the base or furrows at the summit of the guard. (Example—*B. abbreviatus*, Miller. Oolite.)

G. Alveolar cavity very large in proportion, chambered, siphunculated ; no fissure, canal, or grooves. (Example—*B. obtusus*, copied from Knorr, so as to give quite a wrong notion, but it was previously and better figured by Klein (ix, 2). This section is founded in a mistake.)

H. Species incompletely known—a mere appendix.

We have therefore effectively, in Blainville's classification, only five sections of Belemnitidæ, two of which (B and C) may be referred to Belemnitella, leaving three (D, E, F) for the restricted genus of Belemnites.

Blainville adds to his memoir descriptions of Beloptera, Pseudobelus, Rhyncholites, and Conchorhynchus.

In Bronn's 'Lethæa Geognostica,' 1837, we find the Belemnites ranked in three divisions :

A. Integræ.—Sheath without fissure at the basis, with 7—0 furrows at the point. Confined to Lias and Inferior Oolite. Subdivisions according to the number of the furrows near the point.

B. Canaliculatæ.—A distinct canal, beginning from the anterior end, or near it, to or towards the point on the ventral aspect. (Sometimes another canal opposite this.) On the right and left sides often a fine single or double line from the basis to or towards the point. (No ink-bag?) Species mostly confined to the Middle Oolite, but rarely in Coral-Rag, and only peculiar kinds in the Inferior Oolite; two or three in Gault.

C. Fissæ.—Basis with a deep short fissure, not extending backwards beyond the alveolar region; no furrows at the point. Confined to Cretaceous strata.

D'Orbigny separates the third division of Bronn from other Belemnites under the title of Belemnitella; for the others he forms five groups, characterised by peculiarities of the rostrum or guard :[1]

1. Acuarii.—Rostrum more or less conical, often furrowed or wrinkled at the lower extremity, without ventral or lateral furrows in the anterior part. (Lias and Oolites.)
2. Canaliculati.—Rostrum elongate, lanceolate, or conical, provided below with a ventral canal occupying nearly the whole length; no lateral furrows. (Inferior Oolite and Great Oolite.)
3. Hastati.—Rostrum elongate, commonly lanceolate, with lateral furrows on a part of the length, and an anterior ventral furrow, very distinct. (Lias, Oxfordian, Corallian, Neocomian, and Gault.)
4. Clavati.—Rostrum elongate, often clavate, with lateral furrows, but no anterior ventral furrow. (Lias.)
5. Dilatati.—Rostrum compressed, often enlarged, with lateral furrows and a deep *dorsal* anterior groove. (Neocomian.)

This author, from a study of recent Cephalopoda, was led to the supposition, and finally, after researches among fossil groups, to the conviction, that in many Belemnites, independently of age, the rostrum was longer or shorter in proportion to the alveolar axis, according to sex, the males having a longer and the females a shorter guard. But with advancing age, in some cases, the growth of the female guard restores, or nearly restores, the equality. He regards the embryonic Belemnites as always composed of a sphericle and a rostrum with a circular section.

Mr. S. P. Woodward, in 1851,[2] presented a more practical view of the family of the Belemnitidæ than any previous British writer. The definitions of the family and its subdivisions are as under :

BELEMNITIDÆ. *Shell* consisting of a *pen*, terminating posteriorly in a chambered cone (phragmocone); sometimes invested with a fibrous *guard*. The air-cells of the *phragmocone* are connected by a *siphuncle*, close to the ventral side.

[1] 'Terr. Jurassiques,' p. 73.   [2] 'Manual of the Mollusca,' p. 73.

Belemnites.

> *Phragmocone* horny, slightly nacreous, with a minute globular nucleus at the apex; divided internally by numerous concave *septa*. *Pen* represented by two nacreous bands on the dorsal side of the phragmocone, and produced beyond its rim in the form of sword-shaped processes. *Guard* fibrous, often elongated and cylindrical, becoming very thin in front where it covers the phragmocone.

Section I.—Acœli, *Bronn.*

> Without dorsal or ventral grooves.

> *Subsection* 1.—*Acuarii.*

>> Without lateral furrows, but often channelled at the extreme point. (Example—*B. acuarius.* Lias.

> *Subsection* 2.—*Clavati.*

>> With lateral furrows. (Example—*B. clavatus.* Lias.)

Section II.—Gastrocœli, *D'Orb.*

> Ventral groove distinct.

> *Subsection* 1.—*Canaliculati:* no lateral furrows. (Example—*B. canaliculatus.* Inf. Oolite.)

> *Subsection* 2.—*Hastati:* lateral furrows distinct. (Example—*B. hastatus.* Oolite.)[1]

Section III.—Notocœli, *D'Orb.*

>> With a dorsal groove, and furrowed on each side. (Example—*B. dilatatus.* Neocomian.)

Belemnitella, *D'Orb.*

> The guard has a straight fissure on the ventral side of its alveolar border; its surface exhibits distinct vascular impressions. The phragmocone is never preserved, but casts of the alveolus show that it was chambered, that it had a single dorsal ridge, a ventral process passing into the fissure of the guard, and an apical nucleus. (Example—*B. mucronatus.* Chalk.)

---

[1] Blainville, by whom the species was named, quotes it as from Lias, and also from the Clay of the Vaches Noires (Oxfordian). It seems, however, not to be found at all in the Lias.

THE

PALÆONTOGRAPHICAL SOCIETY.

INSTITUTED MDCCCXLVII.

VOLUME FOR 1864.

LONDON:

MDCCCLXVI.

A MONOGRAPH

OF

BRITISH BELEMNITIDÆ.

BY

JOHN PHILLIPS,

M.A. OXON., LL.D. DUBLIN, F.R.S., F.G.S., ETC.,

PROFESSOR OF GEOLOGY IN THE UNIVERSITY OF OXFORD.

PART II,

CONTAINING

PAGES 29—52; PLATES I—VII.

LONDON:

PRINTED FOR THE PALÆONTOGRAPHICAL SOCIETY.

1866.

PRINTED BY J. E. ADLARD, BARTHOLOMEW CLOSE, E.C.

Acanthoteuthis,[1] *Wagner.*

> The shell is not known; probably identical with Belemnosepia. The hooks were in double rows along ten nearly equal arms. (Example—*A. prisca.* Oolite of Solenhofen.)

Belemnoteuthis, *Pearce,* 1842.

> *Shell* consisting of a *phragmocone* like that of the Belemnite, a horny dorsal *pen,* with obscure lateral bands, and a thin fibrous *guard,* with two diverging ridges on the dorsal aspect.
>
> *Animal* provided with *arms* and *tentacula* of nearly equal length, furnished with a double alternating series of horny hooks, from twenty to thirty pairs on each arm; *mantle* free all round; fins large, mediodorsal. (Example—*B. antiqua.* Oxford Clay.)

Conoteuthis, *D'Orb.*

> *Phragmocone* slightly curved. *Pen* elongated, very slender. (Example—*C. Dupiniana.* Neocomian.)

In 1863 Karl Mayer presented, in the 'Journal de Conchyliologie,' the following classification of Jurassic Belemnites, which is illustrated by the collection in the Zurich Museum:

ACUARII. Lanceolate, laterally compressed, without canal or lateral grooves. Lias and Oolite.

> A. Smooth. *AA.* Alveolus excentric.
> *AAA.* Alveolus central.
>
> B. With furrows at the apex.
> *BB.* Alveolus central.
> *BBB.* Alveolus excentric.

CANALICULATI. Lanceolate, with ventral depression, but no lateral grooves. Oolite.

> A. Smooth. *AA.* Alveolus excentric.
> *AAA.* Alveolus central.
>
> B. Unisulcate. *BB.* Alveolus central.
> *BBB.* Alveolus excentric.
>
> C. Bicanaliculate.

---

[1] Professor E. Suess has lately given figures and descriptions of the Acanthoteuthidæ, 'Proceedings of the Imperial Academy,' Vienna, 1865.

HASTATI.   Fusiform, with lateral grooves or canals.   Lias and Oolite.

> A. Smooth.
> B. Unisulcate.
> C. With lateral canals.

Each of the divisions thus indicated is subdivided into groups of allied species.

M. Duval-Jouve proposed a classification limited to the Belemnites of the Neocomian strata, which may be convenient for reference; though at the present time hardly a single species of the groups he mentions and describes so carefully has been found in England.

> 1. Bipartiti.—The sides marked by a deep furrow, which divides them into two equal parts; place of the siphuncle unknown. *B. bipartitus* of Blainville is an example.
> 2. Notosiphiti.—The siphuncle placed on the middle line of the *dorsal aspect* of the alveolus, opposite to the ventral canal of the guard; always compressed; the opening of the cavity notched on the sides.
> 3. Gastrosiphiti.—Siphuncle in the middle line of the *ventral aspect;* always cylindric or depressed; opening of the cavity terminating circularly or obliquely.

Before proceeding further it is desirable to fix the meaning of a few terms of continual use in describing the guard and phragmocone.   In general figure a few Belemnites are very nearly *conical* in the retral part of the guard, having there straight sides and a nearly circular section; more frequently the section is not circular, and the figure is better termed *conoidal.*   Again, in the middle of the guard, as usually found, some Belemnites are very nearly *cylindrical,* with parallel sides and a circular section; but as often other Belemnites have the section oval or in other ways deviating from a circle, and to such the term *cylindroidal* will be applied.   Looked at in a general sense, the whole form of the guard is said to be *hastate* when between the apex of the phragmocone and the termination of the guard the outline is swollen; when this swelling is very slight the term *subhastate* may be employed; when, on the contrary, it is large and conspicuous the Belemnite is called *fusiform.*   Forms which are cylindrical or cylindroidal in the middle of the guard, and conical or conoidal in the hinder part, are often conveniently called *lanceolate.*   As to the termination of the guard, it is in the Belemnitellæ of the Chalk *mucronate;* in some Oolitic and Lias groups this form is nearly approached, and the term *sub-mucronate* will be useful.   In *Belemnitella attenuata* and some others the point is *produced;* and for the remaining forms *acute* and *obtuse,* with the help of the adverbs *very* and *slightly,* will probably suffice.   For want of care in the use of two other terms in respect of the guard, great confusion arises.   Belemnites are *compressed* when the diameter from *side to side* is less than that from the back to the front; *depressed* when the contrary occurs. The *axis of the guard* (called the " apicial line " by Voltz) is the line from the apex of the guard to the apex of the phragmocone; the *axis of the phragmocone* extends from its apex to the centre of the last septum. The relative length of these axes is of much interest and importance in diagnosis. Phragmocones are not always truly *conical;* they are usually a little compressed, and deserve to be called *conoidal.* The angle of inclination of their sides is believed to be nearly constant in the same species, for the same part of the slopes; but it varies in different species ($12°$ to $32°$), and in the same species it often varies a little between the apex and the last chamber. The angle most proper to take for characters of species is between side and side; but it is desirable, when good examples occur, to give also the angle of the back and front. The section of the alveolar chamber may often be had when the phragmocone cannot be obtained.

BELEMNITIC BEDS.

A full account of the geographical and geological distribution of Belemnites must be postponed till the species have been described; but it will be convenient here to indicate the principal zones, or beds of rock, in which they are found most abundantly. Not only is the group absent from existing oceans, but it is unknown in the whole Cænozoic period; for Beloptera and Belemnosis, which occur in Eocene strata, are probably of the family Sepiadæ.

In the Upper Chalk of Kent, Norfolk, and Yorkshire, with *Ananchytes ovata,* we recognise in abundance Belemnitellæ, the latest members of the family in Britain; at Maestricht beds of Chalk, thought to be somewhat higher in the series, also contain them in plenty. Comparatively rare in the lower parts of the Chalk, and not very frequent in Upper Greensand, they are plentiful in the Gault, but again rare in Lower Greensand. No member of the family has been found in the Wealden strata.

Belemnites appear below, but, excepting one dubious notice, I have no information of them in any part of the Purbeck or the Portland strata. They are abundant in Kimmeridge Clay, frequent in Coralline Oolite, less frequent in Calcareous Grit, but again become plentiful in the Oxford Clay, both in the upper part and lower part, as well as in the Kelloway Rock.

Again, they become rare in Cornbrash, and are almost unknown, except as fragments, in Forest-Marble. In Bradford Clay they are unknown to me, except by a notice in Smith's ' Stratigraphical System,' where a small slender species is quoted from Stoford, south of Bradford, in Wiltshire. Nor have I more than mere fragmentary indications in the Great Oolite, till at the base of it we find the canaliculated Belemnites of Stonesfield. Smith mentions a canaliculated species in the Fuller's Earth, but they are very rare in any of the beds between the two Oolites of Bath.

The Inferior Oolite contains many species, in the lower beds especially, and from this point downwards through the sands, and clays, and limestones, of the Liassic series, Belemnites are almost never absent from the section till we reach the zone of *Ammonites Bucklandi*. Only in the upper part of this zone have they been found by Mr. Sanders at Salford, and by Mr. Day at Lyme Regis. I have not, with my own hands, after three careful examinations of the same zone at Lyme Regis, obtained a single example; nor any trace of one, either in the lower parts of that zone or in the subjacent bands with *Ammonites planorbis*. None has ever been seen by me in the strata below from any place in the British Isles.

"Belemnite-Beds" are best exemplified in the Lias, where thin bands of strata are remarkably stored, and even crowded, with the guards of Belemnites. It will be enough to cite in the Southern Lias the well-known rich layer at the foot of Golden Cap, and on the front of Black Venn, near Lyme Regis. On the Yorkshire coast are several of these bands, in the Lower, Middle, and Upper Lias—different species in each of these cases. The Cephalopoda-Bed, as it is called, just at the junction of the Liassic Sands with the Inferior Oolite, is sometimes very rich in Belemnites, and so are parts of the Oxford Clay, the Red Chalk of Speeton, and the upper layers of the White Chalk. As far as mere number is in question, these may be called "Belemnite-Beds," but they are not so in the same sense as the Liassic layers already mentioned.

Till within a short time, the only examples of Belemnites in the strata of Britain, even approaching to completeness, were found in the Oxford Clay near Chippenham, where also shells of Ammonites were more than usually perfect; and other Cephalopoda retained the form of some of the softer parts. Lately, Mr. Day was successful in extracting from the Lias of Lyme Regis several specimens in which the hooks of the arms were preserved, the arms having disappeared, and the greater part of the phragmocone appeared in its place as regards the sparry guard. Possibly, by a careful search in the Gault of Folkstone, the true shape and some further details of the smaller species of Belemnitellæ may be recovered.

The descriptions of species will now be entered upon, beginning with those of the Lias.

BELEMNITES OF THE LIAS.

BELEMNITES ACUTUS, *Miller.* Pl. I, fig. 1.

Reference. *Belemnites acutus*, Miller, ' Geol. Trans.,' 2nd series, vol. ii, p. 60, pl. viii, fig. 9 (read April, 1823), 1826.

B. brevis, var. A, Blainville, ' Mém. sur les Belemnites,' p. 86, pl. iii, figs. 1, 1 *a*, 1827.

B. acutus, Sowerby, ' Min. Conch.,' p. 180, t. 590, fig. 7 (not fig. 10), 1828.

B. acutus, D'Orbigny, ' Terr. Jurass.,' p. 94, t. ix, figs. 8—14, 1842.

B. brevis primus, Quenstedt, ' Cephal.,' p. 395, t. xxiii, fig. 17, 1849.

GUARD. Conoidal, compressed, terminating in a sharp, nearly central point, somewhat drawn out; on each side frequently a broad shallow groove, not reaching to the point; apex often striated, but not grooved.

Transverse section oval, with flattened sides; the ventral aspect broader than the dorsal; axis a little excentric (Pl. I, fig. 1, *s''*).

Greatest length observed, 2·75 inches; greatest diameter, 0·66; axis, 1·00.

Young specimens are longer in proportion, and very acute (Pl. I, fig. 1, J).

Proportions. The normal diameter (*v, d*) at the apex of the phragmocone being taken at 100, the ventral part of it is from 36 to 44, the dorsal 56 to 64; the shorter diameter at the same point is 87; the axis of the guard 300. In young specimens the axis is 400 or more.

PHRAGMOCONE. Oblique, with excentric apex; the *sides* are nearly straight, and inclined at an angle of 27° or 28°;[1] the *angle* included between the dorsal and ventral lines is about 32°. *Section* elliptical, as 100 to 91. *Septa* not observed, except near the apex of the phragmocone.

VARIETIES. *a.* Lateral grooves distinct (specimens figured).

β. Lateral grooves obsolete (not figured).

Observations. The type specimens employed by Miller are unknown. There is in the museum of the Bristol Institution a tablet marked *B. acutus*, Lower Lias, Cheltenham, bearing two specimens. One, dark-gray in colour, corresponds to some of my specimens from the same locality, presented to me by my much regretted friend H. E. Strickland, Esq. The other, composed of yellow spar, fits well enough in figure to Miller's outline, but is of a different species. The figure of this author is quite indeterminate, nor is Sowerby's sufficient for identification. But there is no reason for disturbing the common consent of

[1] D'Orbigny gives the angles as 18° to 24°, probably a mistake.

English and foreign palæontologists, by which the name of *B. acutus* is fixed on the short, compressed, pointed forms here described as containing two principal varieties. Equally general is the consent to adopt as a synonym the first of three varieties, or rather species, ranked by Blainville as *B. brevis;* and by examination of foreign specimens referred to that fossil by M. Hébert, I find specimens of both varieties among them. In the original description of Blainville the axis of the guard is described as medial throughout. Such is, I think, never the case with English examples, though the excentricity varies. D'Orbigny describes the axis as a little excentric. The same author gives a drawing to represent a section of the phragmocone, with sides much less inclined than usual, and with septa unusually distant. I have not yet obtained a good section of a phragmocone. D'Orbigny refers to an angular variety in his own collection, 'Terr. Jur.' (pl. ix, figs. 13, 14). His specimens are a little bent, as sometimes occurs at Lyme Regis and Cheltenham. This author notices the varying length of the guard in proportion to the diameter, the longer specimens being supposed masculine. He did not perceive any grooves on his examples.

Locality and Distribution. In Lower Lias beds, with *Ammonites Bucklandi*, at the base of Black Venn, Lyme Regis (*Day*). In Lower Lias at Weston, near Bath (*Sowerby*), and at Salford (*Sanders*). In Lower Lias beds at Hatch, near Taunton, with *Amm. obtusus* (*Moore*). Antrim, in Lower Lias (*Phillips*). Near Cheltenham, in middle part of Lower Lias (*Strickland*). Robin Hood's Bay, Yorkshire, in upper part of Lower Lias (*Oppel, Phillips*). Thus in the British Islands it seems confined to Lower Lias.

Foreign Localities. In Lower and Middle Lias (communicated by M. Hébert). La Grange aux Bois, Charolles, St. Cyr (Côte d'Or), Argenton. D'Orbigny gives other stations.

BELEMNITES PENICILLATUS, *Sowerby.* Pl. I, fig. 2.

Reference. Belemnites penicillatus, Sow., 'Min. Conch.,' p. 181, t. 590, figs. 6, 9, 1828.

GUARD. Short, much compressed, cylindroid, except in the posterior part, which curves round to an obtuse, nearly central apex; dorso-lateral grooves variable.

Transverse section oval; ventral aspect somewhat broader than the dorsal; axis nearly central.

Greatest length observed, 2·75 inches; greatest diameter, 0·75; axis, 1·00.

Proportions. The longest diameter of the apex of the phragmocone being taken at 100, the ventral part is 48, the dorsal 52; the shorter diameter 76; the axis 150 in old, under 300 in young specimens.

PHRAGMOCONE. Almost truly conical, nearly straight on all sides; its apex almost central; angle 24°; section slightly elliptical, within the oval guard, which is much thickened on the ventral and dorsal faces.

VARIETIES. a. Lateral grooves distinct (fig. 2, b', b''').

β. Lateral grooves obsolete (fig. 2, b''; and Sow., 'Min. Con.,' t. 590, fig. 6).

Observations. In general form this species differs very sensibly from *B. acutus,* as given in these pages; in the transverse section not less so, this having the apex of the phragmocone almost exactly in the centre of the figure, which would be pretty regularly oval but for the flattened sides. Facettes, or grooves, are almost always traceable on the flattened sides, and seldom absent from the apex on the dorsal aspect. They are, however, often obscure enough to escape uncritical observation. With *B. digitalis* it agrees in some degree as to general figure, but as to the proportions of the ventral and dorsal radii of the sheath, and as to the form and situation of the phragmocone, not at all. Very many of the specimens have a rough striation about the apex, arising from some decay there; and, from the same cause, a sort of umbilical depression occurs in place of the original apex, which in the young state was obtuse-angled and entire.

It is necessary to add an additional figure (Diagram 16) for the purpose of preventing mistake when some apparently even-surfaced specimens, with a very symmetrical outline, occur. In most of my drawings especial care is taken to make the groovings fully as evident as they appear in the objects. For this purpose the light is made

DIAGRAM 16.

incident at a lower angle than is usual, so as to mark the longitudinal undulations with distinct light and shade. If this be not done many Belemnites may be regarded as "without grooves," which yet really are furrowed. In the ordinary light such specimens as those in Pl. I, fig. 2, l' and l'', would not appear quite so strongly furrowed, and others, like that in Diagram 16, may be thought to be perfectly smooth.

Belemnites penicillatus is the name given by Blainville to the specimen figured in his work (pl. iii, fig. 1). The name had been previously employed by Schlotheim (' Petrif.,' No. 10), for a Belemnite which Blainville supposes may be the same as his examples, which were from the Lias of Nancy.[1] Hardly any foreign author now employs this name; but it appears desirable to revive it, on the authority of Sowerby, who believed his fossils from Lyme Regis to be the same species as Blainville's. The specimens from the Belemnite-bed of Golden Cap have been sometimes referred to *B. Nodotianus* of D'Orbigny; but that species is represented with a distinct acro-ventral groove, which rarely, if ever, appears in this, and the section given of its phragmocone is very oval, while in this it is almost circular. The two forms are, however, much allied, though not specifically the same. I have seen only one English specimen which appears to agree with *B. Nodotianus*.

Locality. Abundant under Golden Cap, Lyme Regis (*Anning*). Shorn Cliff, Lyme Regis (*Sowerby*). In Lower Lias, with *Amm. Bucklandi,* at Paulton, near Bath, and with *A. obtusus* at Hatch, near Taunton (*Moore*). In Lower Lias (middle part) near Cheltenham (*Strickland*). In Lower Lias, with *A. Turneri,* near Bristol (*Stoddart*). In Lower Lias, Antrim (*Phillips*); and Robin Hood's Bay, Yorkshire (*Phillips*).

BELEMNITES INFUNDIBULUM, n. s. Pl. I, fig. 3.

GUARD. Short, conical, arched upwards; apex acute, usually striated on the dorsal and ventral faces; two obscure lateral facettes extended and widening over the alveolar region. Transverse section nearly circular, with the axis a little excentric, and nearly straight; the young forms similar to the full grown.

DIAGRAM 17.

Greatest length observed, 2·5 inches; greatest diameter, less than 0·7; axis of guard, 0·7.

[1] D'Orbigny twice refers to *B. penicillatus,* once as the equivalent of *B. irregularis,* Sch., and again as a synomym of *B. compressus.*

Proportions. The longest diameter at the apex of the phragmocone being taken at 100, the ventral part is 46, and the dorsal part 54; the axis of the guard 160.

PHRAGMOCONE. Quite straight, with a slightly oval section, ending in a spherule; sides inclined 21°, ventro-dorsal inclination 23°. Septal diameter seven times the depth; axis of the phragmocone half as long as the axis of the guard; alveolar cavity three times as long as the axis of the guard.

Observations. The substance of the guard is usually a clear, brown, calcareous spar, with large radiating pencils of fibres. In some specimens the striations near the apex gather into interrupted longitudinal plaits, especially on the dorsal aspect. In one case the whole surface is marked by little ridgy swellings; in others quite smooth, except near the apex.

Among foreign specimens from Mokon, near Mezières, mostly referred by M. Hébert to *B. brevis*, var. B, Blainville, I have seen some which have the essential characters above assigned, though generally somewhat compressed. Among them one or two may be chosen which fairly match our English examples. Specimens of Lias Belemnites are occasionally seen in foreign museums bearing the names of *B. brevirostris*, D'Orb. (also called *B. curtus* on plate x, figs. 1—6, of this author's 'Terr. Jurassiques,' and *B. rostriformis* (Qu.), which come near to our specimens. The former is known to me by many examples, all of which are straight; the latter is a somewhat oblique, not arched, very short, compressed Belemnite, with acro-lateral grooves, and is ranked among the *Belemnites tripartiti* by Quenstedt. It seems, therefore, that a new designation is necessary for this not common fossil, for which the only synonym I can suggest is *B. brevis*, var. B, in part, of Blainville. It is not identical with what Prof. Morris mentions as *B. brevirostris* from near Cheltenham.

Locality. In Lower Lias, near Bristol, with *Amm. Turneri* (*Stoddart*); Lyme Regis and Bath (*Phillips*). Foot of Black Venn, Lyme Regis, with *A. Bucklandi* (*Day*).

BELEMNITES EXCAVATUS, n. s.   Pl. II, fig. 4.

Reference. Buckland, 'Bridgewater Treatise,' p. 70, pl. xliv', fig. 14, 1836, where it is called *B. brevis* (?), a name previously employed for a very different species by Blainville.

GUARD. Rapidly tapering to a blunt end, with obscure lateral furrows. The axis excessively short. Sections show the substance of the guard everywhere nearly of equal thickness, like the finger of a thick leather glove.

Proportions. The diameter, *v d*, being taken at 100, the axis is of unexampled shortness—less than 100; thus in form somewhat reminding us of *Acanthoteuthis*, though its texture is of the ordinary kind; transverse diameter less than the ventro-dorsal diameter, especially toward the apex.

PHRAGMOCONE. Unknown, excepting that its section was nearly circular, and that its angle was about 28°.

Locality. Lyme Regis, from a calcareous band, probably in the upper part of the Lower Lias. Dr. Buckland's collection one specimen. Professor Phillips's collection, one specimen.

BELEMNITES CALCAR, n. s.   Pl. II, fig. 5.

GUARD. Conoidal, straight, tapering to a blunt apex; sides planate, somewhat inclined to one another; ventral and dorsal surfaces rounded; the ventral aspect broadest. Transverse section oblong, axis very short.

Proportions. The diameter, *v d*, being taken at 100, the axis is of about the same length; the transverse diameter about 90.

PHRAGMOCONE. Only partially known. One specimen of Dr. Buckland's (fig. 5, s'') shows several close displaced septa in the forward part. Its axis must have been five or six times as long as that of the guard. In a specimen belonging to Mr. C. Moore, from Weston, the alveolar parts of the guard are crushed over the alveolar chamber, as in the specimen from Lyme (fig. 5, l', l''); whether they contain any septa can only be known by making sections; and for this more specimens are required.

Locality. Lyme Regis, from the Lower Lias beds, with *Ammonites Bucklandi* (*Geol. Survey* collection, No. 612). Weston, near Bath, in Lower Lias, with *A. Bucklandi* (*Moore's* collection). The specimen Pl. II, fig. 5, l''', is from the Belemnite-beds at the base of the Middle Lias of Lyme Regis. It may possibly be of a different species (*Geol. Survey* collection, No. 613).

Observations. In general figure this Belemnite agrees with *B. brevirostris* of D'Orbigny ('Pal. Fr. Céphalop.,' pl. x, figs. 1—6; on the plate it is called *curtus*), but that species has distinct lateral grooves, which do not appear on the English specimens. It belongs to the Upper Lias of France and Germany; ours as yet appears to be confined to older beds. There is also a resemblance to the incomplete specimen of *Belemnites acuarius macer* figured by Quenstedt ('Cephalop.,' pl. xxv, figs. 27, 28, 29, 30); but those figures are deficient of the extended and striated guard figured on the same plate (pl. xxv, figs. 21, 22). The striations are the effect of the decomposition which has removed the apex. On the same plate fig. 25 represents an individual " der noch die Verlungerung nicht hat," and fig. 26 another; from which it might be supposed that such individuals as ours may be incomplete, and might, indeed, be subject to the same elongation as those of Quenstedt. His specimens are from the uppermost bands of the Lias of Heiningen, ours from the Lower Lias.

BELEMNITES DENS, *Simpson*.   Pl. II, fig. 6.

Reference.   *Belemnites dens*, Simpson, ' Fossils of the Lias of Yorkshire,' 1855 (no figure).

GUARD.   "Length of guard not twice its width, much depressed, sides straight; finely striated longitudinally, or corrugated and roughened most towards the blunt apex, with small tubercles.   Some are longer in proportion."

Locality.   " Lower Lias, Robin Hood's Bay, Yorkshire."

The above is the published description of the curious fossil of which I here present a sketch, in hopes of being able to add more details hereafter.   I have seen only one specimen, which is in the Whitby Museum.   The striations are like those seen on other Belemnites, as *B. elongatus*, and are original marks of structure left by the secreting membrane, or periostracum, not the effect of decomposition, as in *B. acuarius macer* of Quenstedt (' Cephalop.,' pl. xxv, figs. 27, 28, 29).

BELEMNITES CLAVATUS, *Blainville*.   Pl. III, fig. 7.

Reference.   *B. clavatus*, Blainville, p. 97, pl. iii, fig. 12, 1827.
 B. pistilliformis, Sow., ' Min. Conch.,' p. 117, t. 589, fig. 3, 1828.
 B. clavatus, Quenstedt, ' Cephal.,' p. 398, t. xxiii, fig. 19, 1849.
 B. clavatus, D'Orb., ' Pal. Fr. Terr. Jurass.,' p. 103, t. xi, figs. 19—23, 1842.

GUARD.   Very elongate, fusiform, contracted in all the region about the alveolar apex, evenly swollen between this and the apex, which is pointed.   On the lateral faces of the contracted parts of the sheath are traces, more or less distinct, of two longitudinal furrows, which cease on the expanded posterior part, and do not approach the apex.

Sections show the guard to be composed, where it covers the phragmocone, and for a certain space behind the alveolar cavity, of pale, perishable, longitudinal laminæ, which accounts for the frequent absence of those parts, and the production of the form of *Actinocamax*, Mill.   Transverse sections nearly circular, with traces of the grooves about the alveolar region.   With age, the whole figure becomes more lanceolate, and thickens over the alveolar region.

Extreme length observed, 4 inches.

Proportions.   The axis, in ordinary (middle-aged ?) specimens, is from five to ten times as long as the diameter at the alveolar apex.   It is nearer to the ventral side; in some specimens very much so, in others very little.

PHRAGMOCONE.   In a specimen discovered by Mr. Day at Lyme Regis (Pl. II, fig. 7, *s*) the phragmocone is seen in section, and thirty septa are traceable in a length of 0·6 inch. They are converted to iron-pyrites, and seem to be single plates with long flanges. The conotheca is traceable, covered by the thin expansion of the guard. [Quenstedt ('Jura,' p. 137) conjectured that *Orthoceratites elongatus* of De la Beche might be the phragmocone of this species, with its septa very much further apart than is usual in the genus. But that fossil constitutes the *Xiphoteuthis* of Huxley.] The alveolar angle is 18° or 20°, but by compression often appears larger.

Variations. Considerable in respect of the general form and degree of compression of the guard, the excentricity of its axis, and the distinctness of the two lateral furrows in the lower part of the alveolar region. (For notices of these circumstances, see a, β, γ, δ.)

a. Guard with only very faint traces of lateral furrows, and the axis but little excentric.

β. Guard distinctly marked on the alveolar region with two narrow furrows, which vanish on the expanded posterior part. Axis but little excentric. Transverse sections very slightly oval.

Young specimens are elegantly fusiform, and without lateral furrows; with age, laminæ are added over all the surface, so as to elongate the apical region and carry back its swollen part, while in front of this the long, generally furrowed part is of nearly equal diameter. Substance a clear yellowish spar, often white externally, easily decomposing in the contracted part, so as to lose the alveolar portion.

γ. Guard compressed, marked with two lateral furrows; axis excentric, owing to the thickening of the outer layers on the dorsal aspect. This appears to be the *B. spadix ari* of Simpson's 'Lias Fossils,' p. 30.

δ. Guard subcylindrical, without lateral furrows; axis very excentric. (Probably *B. fusteolus*, Simpson, 'Lias Foss.')

The varieties γ and δ are not in general so much contracted in the post-alveolar region as the others.

Observations. Sowerby, who gives good figures, observes, "It is very possibly the young of *Belemnites elongatus*." The remark is not applicable to the fossil which he figures under that name, but there are elongate subcylindrical forms at Lyme Regis, which may perhaps, on further research, be proved to belong to this species grown old. The geological range does not, according to present information, reach the Upper Lias in England.

Localities. In Lower Lias, Hatch, near Taunton, with *Ammonites raricostatus* and *A. obtusus* (*Moore*). In the upper part of the Lower Lias, under Huntcliff, Yorkshire (*Phillips*). In the upper part of the Lower Lias of Robin Hood's Bay, Yorkshire (*Phillips, Cullen*). In the Belemnite-bed at the base of Middle Lias, Golden Cap, Lyme Regis (*Anning, Day, Etheridge, Phillips*). In ironstone-beds east of Staithes, top of Middle Lias, Yorkshire (*Phillips*).

BELEMNITES COMPRESSUS, *Stahl.* Pl. III, fig. 8.

> *Reference.* *Belemnites compressus,* Stahl, ' Correspondenzblatt der Würt. Landw.
> Vereins,' pl. xxxiii, fig. 4, 1824.
> *B. Fournelianus,* D'Orb., ' Terr. Jur.', p. 97, pl. x, figs. 7—14, 1842.
> *B. compressus,* Quenstedt, ' Cephal.,' p. 405, pl. xxiv, figs. 18, 19, 1849.

GUARD. Much compressed, expanded posteriorly, and ending in an obtuse or rounded apex; anteriorly contracted, and sometimes quadrate over the alveolar region. In some specimens the obtuse point is replaced by a cavity plaited at the edges, as in *B. umbilicatus* of Blainville. From the apex a short broad groove proceeds nearly along the middle of each side, and extends towards the alveolar region. On that region one distinct, narrow, dorso-lateral groove appears, and extends towards the apex; a short broad groove is also seen in some specimens on the anterior part of the ventral face of the alveolar region.

Transverse sections show an oval outline and a nearly central axis.

Greatest length observed, 2 inches; but it grows to greater length. (Quenstedt, ' Cephal.,' pl. xxiv, fig. 18.)

Proportions. The diameter, $v\,d$, at the apex of the alveolus being taken at 100, the ventral part is 48, the dorsal 52, the transverse diameter 75, and the axis about 300. This proportion of the axis varies; the individuals with longest axis are supposed by D'Orbingy to be males.

PHRAGMOCONE unknown. The alveolar angle is 28°.

Variations. Considerable, in regard to the completeness of the grooves and striations and the degree of expansion near the apex. A curious monstrosity occurs among Mr. Moore's specimens from Ilminster. In some specimens from Lyme two parallel narrow grooves appear on each side; in others one broad shallow groove only.

Locality. Lyme Regis, in Belemnite-beds of Middle Lias (*Geol. Survey*). Glastonbury, in Middle Lias (*Oxford Museum*). Ilminster, Middle Lias (*Moore*). I have seen no specimen on the coast of Yorkshire.

BELEMNITES BREVIFORMIS, *Voltz.* Pl. IV, figs. 9 A, 9 B, 10A, 10 B, 10 C, 10 D.

> *Reference.* *B. breviformis,* Voltz, ' Obser. s. Bélem.,' p. 43, pl. ii, figs. 2, 3, 4, 1830.
> *B. breviformis amalthei,* Quenstedt, ' Cephal.,' p. 405, pl. xxiv, figs. 21—23,
> 1849.

GUARD. Short, lanceolate, cylindroidal, subtetragonal, tapering to a pointed summit, which is more or less recurved towards the back.

Longitudinal sections show the axis to be always more or less curved, especially near the apex, and nearer to the ventral than the dorsal face; much nearer towards the apex. Transverse sections show a tendency to flatness of the sides, so as to give an approach to a certain squareness in the outline. Diameters nearly equal.

Size rarely exceeding 2 inches; diameter rarely exceeding ⅔ of an inch.

Proportions. The normal diameter being 100, the ventral part of it is 37, the dorsal part 63; the transverse diameter is also 100 (more or less). The axis, 200 to 300.

PHRAGMOCONE. Oblique, with a circular section, more or less inflected towards the apex, and terminated by a sphericle. Concamerations numerous; siphon not affecting the sutures of the cells. Angle 25° to 27°.

VARIETIES. According to M. Voltz, who described this species after inspection of more than fifty individuals from Gundershofen, in Upper Lias, three varieties occur:

(A) Guard somewhat depressed, its summit acutely conical, without distinct inflexion or furrows (Voltz, pl. ii, fig. 2).

(B) Summit submucronate (Voltz, pl. ii, fig. 4).

(c) Summit mucronate, axis very near the ventral side (Voltz, pl. ii, fig. 3).

Observations. The species called *breviformis* by Voltz, if allowed to include all the varieties which have been referred to it, must be quoted from Middle Lias, Upper Lias, and Inferior Oolite. Voltz supposes that *Belemnites brevis*, var. B, of Blainville (pl. iii, fig. 2), may be identical with *B. breviformis*, var. c, Voltz; and as far as the figure of Blainville is evidence, his opinion seemed just. But by late researches of M. Hébert, who has examined many specimens of the variety mentioned by Blainville, it seems to be really a distinct species, for which the name of *brevis* may be retained. This author is of opinion that *B. brevis*, D'Orb. ('Pal. Franç., pl. ix, figs. 1—7; in the text it is called *B. abbreviatus*), is identical with the species of Voltz.

VARIETIES. I possess some half-dozen individuals of this species from Lyme Regis, some of which were collected by myself from the Belemnite-bed, under Golden Cap; others supplied by Miss Anning, probably from the same locality. From Gundershofen M. Voltz was so good as to send me five specimens, showing the varieties A, B, c, which he includes in the species. M. Hébert also sent me a larger series from the same place. It is evident that these all correspond, the English specimens being chiefly of the first variety (A) of Voltz. All agree in a lanceolate figure, with an approach to cylindrical section (or a little planate on the sides), and in a summit quite free from systematic grooves. The apex tends to recurvation (and in var. B, c of Voltz to submucronation). The points of doubt which arise on comparison of specimens from England and Germany are unimportant, but there are peculiarities in the description of M. Voltz which seem to require attention. He notices in his variety A the phragmocone as sensibly curved towards the ventral side; in a section of a specimen from Gundershofen I find this curvature, as (indeed his figures (pl. ii, D″ and D‴) show it to be) very slight. He mentions also a curvature of the axis of the guard, and it appears in the figures quoted; it is only just traceable in my specimen. There is no sensible change of figure from youth to age, except that the diameter grows larger in proportion.

In his variety B (Voltz, fig. 4, D) he figures the axis as decidedly curved, and passing very near to the ventral face in all the posterior parts; the phragmocone is represented as very sensibly curved; neither in B nor C are either ventral or lateral grooves. These varieties are not larger than var. A. In specimens which I possess from the Eston Nab Ironstone-beds, Prees, and Glastonbury, these characters are found precisely as in Voltz's figure and description, except as to the lateral grooves, which are traceable, more or less distinctly, in most cases.' These individuals are larger than any of the specimens of Voltz, and correspond in magnitude with others from Alderton, in Gloucestershire, in which the lateral grooves are quite distinct. Admitting all these forms to belong to the species so well examined by M. Voltz, we have the following result for the British deposits :

VAR. a = var. A, Voltz. Guard having an acute conical termination, without distinct inflexion or grooves. Phragmocone very slightly incurved (Pl. IV, fig. 9 A, B).

Locality. In England, the Belemnite-bed under Golden Cap, Lyme Regis, base of Middle Lias; in Germany, Gundershofen, Upper Lias.

VAR. β = Var. B, C, Voltz. Guard terminated by a summit more or less prominent, acute, and inflected towards the back; no distinct lateral grooves. Phragmocone distinctly incurved (Pl. IV, fig. 10 A).

Locality. In England, Prees, Salop, in Middle Lias (*Morton*). Eston Nab, Yorkshire, in ironstone of the Middle Lias (*Phillips*). Glastonbury, in Middle Lias (*Phillips*).

VAR. γ. Guard terminated by a submucronate summit, more or less prominent from the ventral half of the substance, and more or less inflected towards the back. Lateral furrows near the back always traceable (Pl. IV, fig. 10 B, 10 C, 10 D).

Greatest length observed, 2·5 inches; greatest diameter, 0·6 inch.

Longitudinal sections show the axis to be excentric, arched, and much nearer to the ventral side, remarkably so at the apex of the phragmocone, least so towards the apex of the guard. Transverse sections show the sides to be flattened, so that the sparry substance is thinner there over the alveolus. The lateral grooves at the apex are continued more or less distinctly into these flattened spaces.

Phragmocone arched, ending in a spherule; angle 27° in the anterior part; nearly 30° towards the apex. Transverse section nearly circular. Septa frequent.

If it be found eventually desirable to separate these varieties, it will be best, I conceive, to make the division so as to insulate the variety *a*, which appears to me rather indeterminate, and to resemble the young of other species too much to be quite satisfactorily identified among the Lyme specimens.

Localities. Alderton, Gloucestershire, in marlstone (*Phillips*). Bredon Hill, marlstone (*Strickland*). Eston Nab, and East of Staithes, Yorkshire, in ironstone above the marlstone (*Phillips*). Ilminster, in marlstone of Middle Lias (*Moore*); this is shorter and stouter than usual.

Synonyms. D'Orbigny refers the fossils figured by him pl. ix, figs. 1—7, to this species, expressly referring to Voltz, pl. ii, figs. 2, 3, 4. He calls it in the text *B. abbreviatus* of Miller and Sowerby; on the plate and in the text it is entitled *B. brevis*, Blainville. Authors who quote D'Orbigny in the 'Pal. Franç.' should observe the difference of the names on the plates and in the text. The specimen figured by D'Orbigny does not well agree with Voltz's species; it is larger, has no trace of lateral furrows, and according to the drawing a straight-sided phragmocone, with a decidedly oval section. That it was one of the varieties included by Miller in his *B. abbreviatus* may be readily supposed. It is not the exact equivalent of *B. breviformis amalthei* of Quenstedt (pl. xxiv, figs. 21—23), which has furrows near the apex, but is rather comparable to *B. breviformis*, Quenstedt 'Cephal.,' pl. xxvii, figs. 22—27; (*B. Gingensis* of Oppel), an Oolitic rather than Liassic form, of the South of England, which will be noticed immediately.

BELEMNITES GINGENSIS, *Oppel.*   Pl. V, fig. 11.

Reference.   *Belemnites breviformis*, Quenstedt, 'Cephalop.,' p. 428, t. xxvii, figs. 23—26, 1849.
            B. Gingensis, Oppel, 'Jura,' p. 362, 1856.

Guard. Short, conoidal, contracted, and curving rapidly to an acute, produced, submucronate, rather recurved summit; no distinct grooves about the summit; no distinct flattening of the sides.

Sections show the general outline nearly circular, the axis very excentric and arched, and very near the ventral face.

Greatest length observed (the edge being very thin) under $2\frac{1}{3}$ inches; greatest diameter, under $\frac{3}{4}$.

Proportions. The diameter at the apex of the phragmocone being 100, the ventral radius is about 30, the dorsal 70; the axis varies between 160 and 210; the diameters are nearly equal.

Phragmocone. Incurvate, with an angle of 28°, and a nearly circular section. Septa numerous, distant from each other about ⅛th of the diameter.   Siphuncle marginal, moniliform, quite free from the external layer of the conotheca, and completely bordered by its own shell (a single plate?).   The flanges are short.

Locality.   In the Inferior Oolite of Dundry Hill, with *Ammonites Humphreysianus* (*Bristol Museum*).   In the same beds at Wotton-under-Edge, Frocester Hill, Cam Down, Bridport (*Phillips*).   Near Cheltenham (*Buckman*).

Observations.   The specimens figured are, without doubt, to be referred to the species first separated from the ordinary *breviformis* by Oppel.   I have seen foreign examples. The figures given by Quenstedt show some degree of lateral compression, but I cannot doubt the identity of his species with ours from Gloucestershire and Somersetshire. D'Orbigny's figures ('Terr. Jur.,' pl. ix, figs. 1 and 3), which are said to be from the Upper Lias, agree with our specimens well enough, but the transverse section of the phragmocone (fig. 4) is so remarkably oval that, if not due to compression, there must be some mistake.   The phragmocone is represented in fig. 2 of the same plate, as quite straight on the ventral side.

BELEMNITES INSCULPTUS, n. s.   Pl. V, figs. 12, 13.

Reference.   (I can find no satisfactory figure or description of this Belemnite.)

GUARD.   Short, conoidal, tapering rapidly to a produced, submucronate, somewhat recurved summit, from which two broad, shallow, lateral furrows proceed, growing less distinct over the alveolar region.

Transverse sections show the outline to be nearly circular, but flattened more or less on the sides; the axis very excentric towards the ventral face, and arched.   Sometimes the dorsal aspect is widest, sometimes the ventral; it is flattened sometimes on the ventral face, as in *B. ventroplanus* of Voltz.

A section taken lengthways shows the axis to be excentric, and arched in accordance with the reflected apex.

Greatest length observed (edge of aperture very thin), 3½ inches; greatest breadth, less than 1 inch.

Proportions.   The diameter at the apex of the phragmocone being 100, the ventral radius is 42, the dorsal 58, the axis 120.

PHRAGMOCONE.   Very oblique, arched, with a circular cross section.   Conotheca concentrically undulated on the ventral face; concamerations numerous; angle m. 28°. Septal diameter eight or even nine times the depth.   Axis of phragmocone traced to twice the length of the axis of the guard.   Fifty septa in three quarters of an inch from the apex of the phragmocone.

Locality.   Inferior Oolite, with *Ammonites Humphreysianus*, Dundry (*Sanders*), fig. 12.

7

In the Lias of Lyme Regis occurs very rarely an allied form represented in fig. 13, from specimens in the cabinet of Mr. Goodhall, inspected many years since. I only know the exterior of these examples. A longitudinal section of another specimen from

DIAGRAM 18.

Lyme, preserved in the collection of the Bristol Institution, shows how short is the axis of the guard (Diagram 18). This Belemnite is not mucronate. If additional specimens come to my hands, I hope to determine its characters.

BELEMNITES LATISULCATUS, n. s.   Pl. V, fig. 14.

Reference.   (I can find no satisfactory figure or description of this Belemnite.)

GUARD.   Short conoidal, a little compressed, tapering rapidly in a curve on the ventral side to an obtuse recurved summit, from which two broad, distinct, dorso-lateral grooves proceed, widening over the alveolar region, and margined on the dorsal side by a distinct ridge.

Sections.   The ventral aspect somewhat wider than the dorsal.

Proportions.   The diameter ($v\,d$) at the apex of the phragmocone being 100, the ventral radius is about 45, the dorsal 55, the axis 180, the transverse diameter 94.

PHRAGMOCONE.   Not distinctly observed.

Locality.   Upper Lias of Whitby, the specimen figured (*Phillips*). In the Museum of the Yorkshire Phil. Society is a specimen resembling this in general figure, but more slender, and with only short dorso-lateral grooves.

Belemnites paxillosus, *Schlotheim.* Pl. VI, fig. 15.

Reference. *Belemnites paxillosus,* in part, Schlotheim, 'Taschenb.,' pp. 51 and 70, 1813.

        „   „   Voltz, 'Bélemnites,' p. 50, pl. vi, fig. 2, and pl. vii, fig. 2, 1830.

        „   „   Zieten, 'Wurtemb.,' p. 29, pl. xxiii, fig. 1, 1830.

B. Bruguierianus, D'Orb., 'Pal. Fr. Terr. Jurass.,' p. 84, t. vii, figs. 1—5. The plate is really numbered pl. vi, and the name on it is *Belemnites niger,* also given on pl. v to another species. 1842.

B. paxillosus amalthei, Quenstedt, 'Cephal.,' p. 402, pl. xxiv, 1, 2, 4—8, 1849.

B. paxillosus, Oppel, ' Jura,' p. 152, 1856.

GUARD. Smooth, elongate, cylindroidal, convexo-conical towards the summit, which is often subtruncate or even concave, and marked by two short latero-dorsal smooth furrows, and in most cases by one or more medio-dorsal striæ, and sometimes one medio-ventral short stria.

Sections show the axis to be nearly straight in all the young forms, only bending near the apex in the older examples. The axis is a little excentric.

Greatest length of axis, $3\frac{1}{2}$ inches, and of whole guard, $6\frac{1}{4}$ inches; diameter at apex of alveolus not exceeding 1 inch.

The young have nearly the same general figure and proportions as the adult individuals, and exhibit the same diversities as to compression, and sometimes assume a subhastate shape. In very young forms the proportion of the axis of the guard is found to be somewhat less than in those of middle age. Thus in a specimen from Wurtemburg the proportion in the youngest guards is 260, but in the same full-grown individual 380.

Proportions. The diameter (*v d*) at the apex being 100, the ventral radius is 40 to 45, the dorsal 55 to 60, the axis about 350 to 450.

PHRAGMOCONE. Straight (or very nearly so), with a nearly circular section, the sides meeting at an angle of 22° to 24°; conotheca distinctly striated, the straight striæ bifurcate toward the ventral region (Voltz); chambers rather shallow, numerous, their septa almost directly transverse. (See figures of great value in Voltz, pl. vii, fig. 2.) The depth of the chambers is one seventh of their diameter. The axis ends in a spherule.

In a specimen from Ilminster the septa are traced in section through a length on the axis of the phragmocone equal to half that of the axis of the guard. These septa are to be counted to above 60 in a space of $1\frac{1}{2}$ inch, the anterior septa being broken off from their flanged extremities, and lost; the flanged extremities remain on the ventral side. Broken septa occur within others which are not broken.

Localities. In Middle Lias (marlstone), at Tilton-on-the-Hill, Liecestershire; Cay-thorp, in Lincolnshire, Belvoir Castle, and Oakham (*Phillips*). In Middle Lias (marl-stone), Ilminster (*Moore*). In Middle Lias (marlstone), Staithes, Yorkshire (*Phillips*).

Observations. The history of this interesting Belemnite is contained in specimens from the marlstone and Upper Lias, to which, as far as yet appears, its geological period is limited. Voltz, indeed, quotes a specimen from Oolite near Caen. The British localities are almost exclusively in marlstone; the foreign stations include Upper Lias, in Wurtemburg. Variations are sensible in the guard, in regard to the transverse section, which is sometimes a little compressed; the general figure, which is occasionally a little swollen towards the apex, and sometimes bent upwards; and the terminal grooves and striæ, which latter on the dorsal and ventral faces are sometimes distinct and sometimes wholly wanting. Occasionally there are several of these, and in this case they may be due to decomposition of the laminæ. The lateral grooves are always short, and usually very distinct. In the interior the axis of the guard is more or less excentric than the proportions given above. The sides of the phragmocone, if not quite straight, as Voltz

DIAGRAM 19.

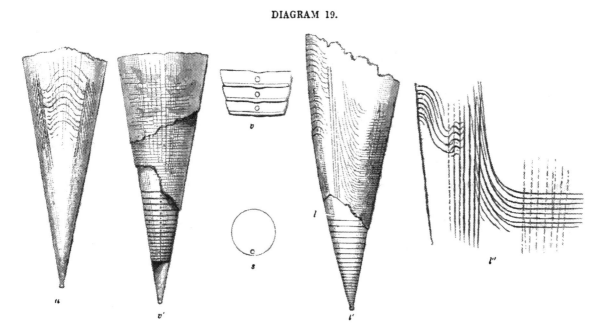

affirms, are nearer to straightness than is usual when the axis is excentric. In D'Orbigny's fig. 5, the phragmocone is more incurved than is at all usual. Natural sections lengthways of this Belemnite are rather frequent in the ventro-dorsal plane, which, near the apex, is sometimes marked by a real fissure.

Mr. Moore's large collection of these Ilminster Belemnites contains phragmocones in different conditions of exposure, from which several facts of interest can be gathered, as the accompanying diagrams (19, 20) may show.

DIAGRAM 20.

In the first place, some traces of growth lines appear on the alveolar cavity, and thin layers adhering to the conotheca, which indicate the terminal edge of the guard to have been in the form seen sideways at *l*, dorsally at *d*, and in front at *v* (Diagram 20). Next the conothecal surface itself shows the structure lines; the dorsal aspect as in *d'*, the side aspect as in *l'*, enlarged in *l''*; and the ventral aspect as in *v'* (Diagram 19). The points to remark are the bifurcation of the arched side lines in *l''* before passing over the front in *v''*; the numerous transverse striæ in *v'* are from 6 to 8 to each concameration; and the undulated outline which the phragmocone deprived of conotheca presents (Diagram 19 *v'*) The transverse striæ in *v'* are not so numerous in the smaller parts of the conotheca; there are only very faint longitudinal medio-dorsal lines; the siphuncle makes no distinct appearance externally. The sides of the phragmocone are nearly straight; it is terminated by. a distinct and rather large spherule. The angle is about 20° (19° to 23°); the cross section *s* nearly circular. The excentricity of the alveolar apex variable, in some specimens very slight. A gentle retral wave in each septum on the siphuncular line.

BELEMNITES APICICURVATUS, *Blainville.* Pl. VI, fig. 16.

> *Reference.* *Belemnites apicicurvatus*, Blainville, 'Mém. sur les Bélemnites,' p. 76, pl. ii, fig. 6, 1827.

GUARD. Elongate, compressed, smooth. Alveolar region compressed, expanding towards the aperture; apicial region convexo-conical, obliquely inclined or even bent towards the back, marked by two latero-dorsal furrows, extending a short distance from the summit, and two still shorter, terminal, latero-ventral grooves. Between these four grooves are often sharp short plaits. In very perfect specimens the whole apicial region is minutely ornamented by longitudinal striations, which near the apex are straight and even continuous, but elsewhere undulate and anastomose. The anterior part of the alveolar region is sometimes roughly striated. Of the plaits at the summit, the medio-ventralis often very short and sharp.

Sections show the apical line almost straight, nearer to the ventral face; the successively superposed laminæ of growth very distinct, the fibres faint, oblique, and somewhat curved near the apex; the substance honey-yellow spar. The ventral region is generally broadest, the sides of the alveolar portion are usually flattened. This species sometimes cracks naturally along the lateral faces.

Greatest length observed, 5·5 inches ; greatest diameter, 0·85 ; axis of guard, 4 inches.

Proportions. The diameter (*v d*) at the alveolar apex being 100, the ventral radius is 37 to 40, the dorsal radius 60 to 63, the diameter from side to side 88, axis 450 to 600.

PHRAGMOCONE. Oblique, incurved ventrally, with oblique septa, and an elliptically compressed section ; the dorsal and ventral faces (curved) inclined at an angle of 29°, the lateral faces (straight) at 25°. On the alveolar shell the dorsal region is defined by a somewhat prominent longitudinal line. If the diameter (*v d*) of the largest chamber be taken at 100, the diameter from side to side = 90, the depth of the chamber = 12, and the longest side of the alveolus 300. The axis of the phragmocone = half the axis of the guard.

Localities. The Belemnite-bed, base of Middle Lias, at foot of Golden Cap, Lyme Regis (*Miss Anning*). Cheltenham (*Strickland*).

Observations. Described by Brongniart as from Lyme Regis in 1826. It has been referred by D'Orbigny to *B. compressus* of Blainville, but its true affinity is to *B. elongatus,* as given by Sowerby, and to *B. paxillosus* of Voltz. It varies as to the degree of lateral compression and as to the terminal plaits and striæ.

BELEMNITES ELONGATUS, *Sowerby.*   Pl. VII, fig. 17.

Reference.   Sowerby, 'Min. Conch.,' p. 178, t. 590, fig. 1, 1828.
Quenstedt, 'Cephal.,' p. 402, t. 24, fig. 3, 1849.
Huxley, in 'Memoirs of the Geological Survey,' Monograph II, pl. i, figs. 2, 3, 1864.

GUARD. Rather compressed, cylindroidal about the apex of the phragmocone, thence tapering in regular sweep to the acute-angled point, and expanded towards the aperture ; dorso-lateral grooves distinct for only one fourth or one third of the length of the axis, thence obscurely prolonged into the somewhat flattened sides ; small plaits, striations, and granulations on the surface, especially near the apex ; no ventral furrow ; axis a little excentric, most so at the apex of the phragmocone.

Sections show the axis excentric, within a slightly oval outline, and nearly straight.

The largest individual yet observed is the fine specimen figured by Sowerby, and now preserved in the British Museum, which is complete from the apex to the last or nearly the last septum of the chambered cone, and measures 10½ inches in length. Of this the axis of the guard is about 3 inches, that of the phragmocone nearly 8 inches ; longest diameter of section at the apex of phragmocone, 0·9 inch. My smallest specimen from Lyme Regis measures 0·38 inch diameter, and 1·65 long.

In this young individual the lateral grooves are quite distinct, and continued to the acutely tapered apex ; two striæ appear on the dorsal aspect at the apex, and one small plait on the ventral aspect.

Proportions. Taking the dorso-ventral diameter at 100, the ventral radius is 40, the dorsal radius 60, the transverse diameter, 94; the axis of the guard is about 300, but is more or less according to age and range of natural variation. In my smallest specimen the axis of the guard is 345.

PHRAGMOCONE. Excentric; transverse section a little oblong, longitudinal section a little incurved, angle 25°.

Localities. Crick Tunnel, specimen figured (*Sowerby*). Belemnite-bed, Lyme (*Day, Etheridge*). Cheltenham (*Strickland*). I have not seen this species in the strata of the Yorkshire coast.

Observations. There is no certain knowledge of the specimens which served Mr. Miller for the type of this species, nor is the figure which he gives, or the description which accompanies it, at all critical. In fact, they have been interpreted so variously, and referred to so many different species, as to be of little or no authority. No grooves are mentioned on the guard, which is merely described as "slender, tapering to a conical point." The figure of Mr. Sowerby is taken from a fine specimen now in the British Museum, and gives a good general representation of the fossil. The description, however, is not only incomplete, but inexact on an important point. "Slender, cylindrical in the middle, gradually expanding to a broad base one way, and tapering to a point the other; round, and *free from furrows*; the chambered cavity two thirds of the length of the shell." The localities quoted are Charmouth, Bath, Crick Tunnel, all in the Lias. Instead of being "free from furrows," the specimen has the usual two dorso-lateral grooves near the apex clearly defined; on the ventral aspect at the apex is a little elevated plait, but no furrow. By these characters it belongs to the natural group of the "*paxillosi*;" the section of the guard is not quite round, even in old specimens, but always a little compressed, often sensibly so when young, with an excentric axis; the apex is more acute than is usual with *B. paxillosus*, and the phragmocone is very much more extended than in that species.

The figure given in the monograph of the Geological Survey, already referred to, agrees with the specimen in the British Museum and with several in my own collection sent me from the "Belemnite Bed" of Lyme Regis by Miss Anning. For the drawings on Plate VII, which represent the original specimen in the hands of Sowerby, I am indebted to Mr. H. Woodward. Adopting the specimens preserved in the British Museum and in Jermyn Street for types of the adult, I have attempted, chiefly by help of the specimens in my own drawers, to trace the younger forms, and determine the limits of variation to which the species is subject; but I am not sufficiently provided with specimens, especially for sections.

In a specimen from Upper Lias, in my collection, which has the distinct lateral grooves and the external characters of *B. elongatus*, there is a small, narrow, deep, stria on the ventral aspect, close to the apex, and another on the opposite dorsal aspect. In another example, also from the Upper Lias, and in my collection, both the dorsal and ventral surfaces are distinctly marked with one stria from the apex, so that if these be regarded as furrows, the apicial region is quadri-sulcate.   The apices are in each case more pointed and more recurved than is usual with *B. paxillosus*.   A longitudinal section of the former specimen shows the axis of the guard most excentric at the apex of the phragmocone, and the youngest (included) forms to have been relatively much shorter than the older.

In the accompanying sketches (Diagram 21) the smaller one shows the proportions of the sheath in the youngest traceable form of this specimen ; the larger shows a middle-age

DIAGRAM 21.

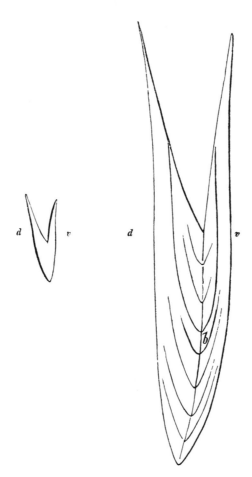

form (*b*) included in a third a full-grown specimen, whose probable extension is given. In the three cases the proportion of the axis of the guard to the normal diameter (*v, d*), taken at 100, is 140 for the youngest, 260 for the middle-age, 300 for the full-grown. The proportion of the whole length of the guard to the contemporaneous length of the axis is in the first case 260, in the second 180, and in the third (inferred, not measured) it is nearly the same.   In the first case the length of the whole guard is to its **greatest** diameter about as 200 to 100, in the second about 350 to 100, and this proportion is not materially altered with further growth.

EXPLANATION OF PLATE I.

FIG.

1. BELEMNITES ACUTUS.

d. Dorsal aspect.

l. Lateral aspect. The groove here seen is seldom so distinct.

v. Ventral aspect.

s. Section from the back to the front. The alveolus is excentric, and is expanded by compression : *v*, ventral; *d*, dorsal portion.

J, J. Young specimens, showing their elongate form.

a. Apicial portion of the guard, showing striation, which is not usual, and is referable to decay.

s″. Transverse section at the apex of the alveolus, showing the greater breadth of the ventral region, which, however, is here somewhat broader than usual.

2. BELEMNITES PENICILLATUS.

d. Dorsal aspect.

l′, l″, l‴. Lateral aspect; *l′* and *l‴* show lateral grooves and terminal striæ. These grooves are seldom quite untraceable, though they do not appear in *l″*, a younger specimen, which, however, is more striated, and at the apex umbilicate; these appearances in *l″* are due to partial decay.

v. Ventral aspect.

s. Longitudinal section, showing the straight-sided central alveolus.

s′ Transverse section, showing the elliptical section of the alveolus and the flattened sides of the guard.

s″ Transverse section of the guard, showing the central axis.

3. BELEMNITES INFUNDIBULUM.

d′, d″, d‴. Dorsal aspect, showing the usual striæ, which are enlarged in σ.

l′, l″. Lateral aspect, showing the usual inflexion of the apex.

v. Ventral aspect, on which the striæ are usually shorter. These striæ are not due to decay, but to original formation.

s″ Transverse section, nearly round, or subquadrate, or a little oval, according to the specimen and the place of the section.

σ. Enlarged striæ.

PL. I.

Fig. 1.

Fig. 2.

Fig. 3.

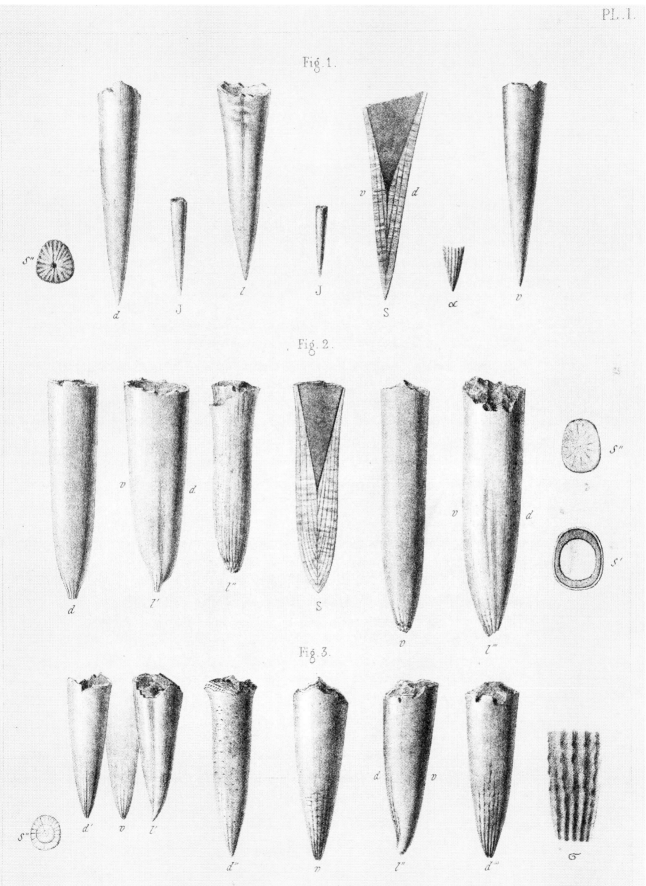

P. Lackerbauer lith.

Imp. Becquet, Paris.

EXPLANATION OF PLATE II.

Fig.

4. BELEMNITES EXCAVATUS.

  v. Ventral aspect. (Prof. Phillips's specimen.)

  l. Lateral aspect; *d* and *v* mark the dorsal and ventral portions. The apex of the alveolar cavity is between those letters. (Prof. Phillips's specimen.)

  s. Longitudinal section, showing the very thin sheath and very deep and ample alveolar cavity. (Dr. Buckland's specimen.)

  s′. Transverse section across the alveolar cavity and guard. (Prof. Phillips's specimen.)

  s″ Transverse section of the same specimen near the apex.

5. BELEMNITES CALCAR.

  l′, l″. Lateral views of the specimen belonging to the Geological Survey (No. 612).

  d. Dorsal view of the same.

  s. Transverse section of the same.

  s. A rather oblique section across the alveolar cavity, in consequence of which the sheath appears thicker towards the apex than it would appear on a truly axial section. (Oxford Museum.)

  l‴ Lateral view of a specimen supposed to be of this species. Geological Survey Collection (No. 613).

6. BELEMNITES DENS.

  l. General figure, seen laterally. The specimen is compressed, as may be seen by the transverse sectional outline, *s′*.

  The striæ are seen magnified at *σ′* and *σ″*.

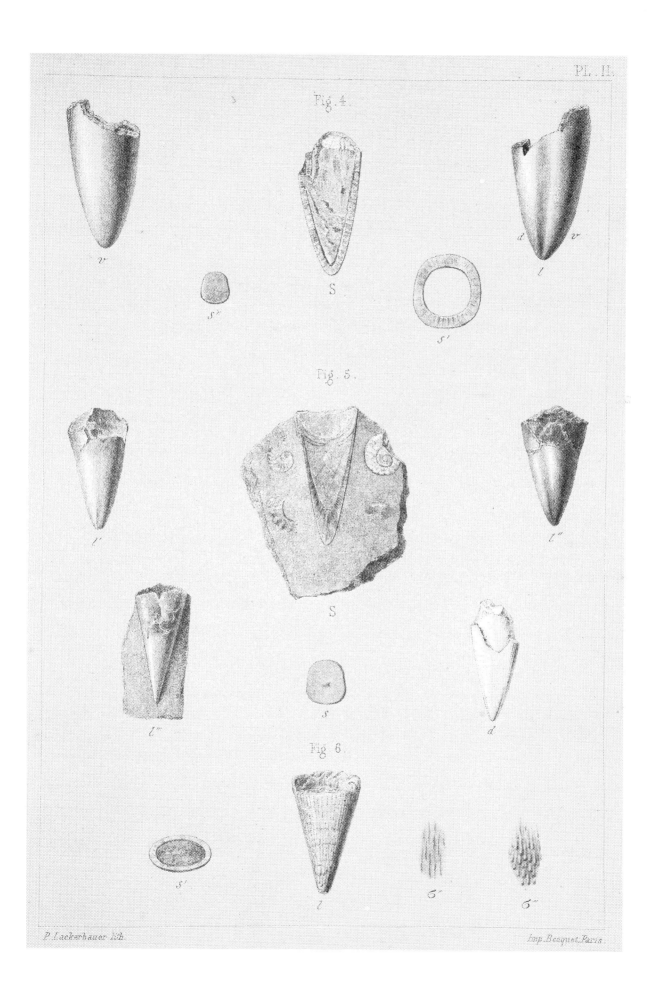

Fig. 4.

Fig. 5.

Fig 6.

P. Lackerbauer lith.

Imp. Becquet, Paris.

EXPLANATION OF PLATE III.

Fig. 7.

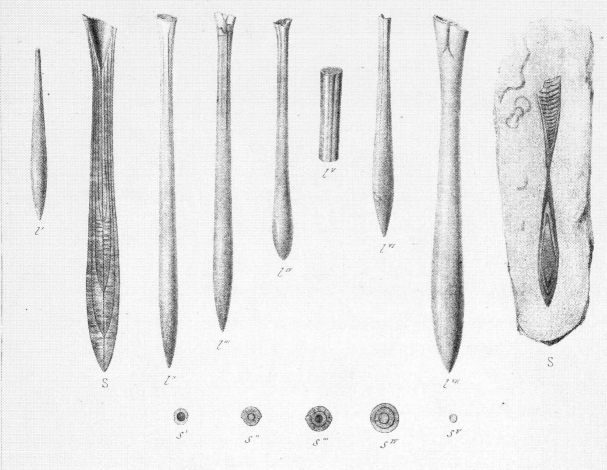

Fig. 8.

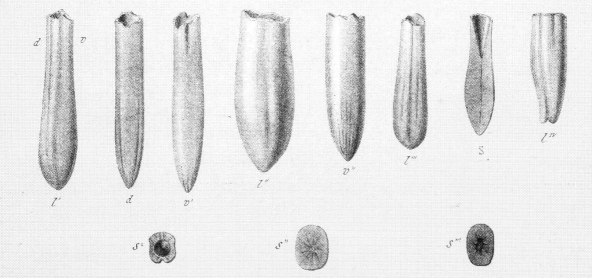

P. Lackerbauer lith.

Imp Becquet, Paris.

EXPLANATION OF PLATE IV.

F<small>IG.</small>

9A. B<small>ELEMNITES BREVIFORMIS</small>, *var. a* (*var.* A, *Voltz*), from Lyme Regis.

  v. Ventral aspect.
  l. Lateral aspect.
  s′. Transverse section.

9B. B<small>ELEMNITES BREVIFORMIS</small>, *var. a* (*var.* A, *Voltz*), from Gundershofen.

  v. Ventral aspect.
  l. Lateral aspect.
  s′. Longitudinal section, showing a nearly straight axis of guard, and a spherule at the apex of the phragmocone.
  s″. Section to show the form of a very young individual.
  s‴. Shows the spherule.
  s⁗. The terminal laminæ of the guard.
  s′. Transverse section, showing a slight ventral flattening.

10A. B<small>ELEMNITES BREVIFORMIS</small>, *var. β* (*var.* B, C, *Voltz*), from Prees, Salop.

  d. Dorsal aspect.
  l. Lateral aspect.
  v. Ventral aspect.

10B. B<small>ELEMNITES BREVIFORMIS</small>, *var. γ*, from Ilminster.

  d. Dorsal aspect.
  l. Lateral aspect.
  s′. Section across the alveolus.
  s″. Section across the guard, near the apex.

10C. B<small>ELEMNITES BREVIFORMIS</small>, *var. γ*, from Alderton.

  d. Dorsal aspect.
  v. Ventral aspect.
  l. Lateral aspect.
  s. Longitudinal section.
  s′, s″. Transverse sections.

10D. B<small>ELEMNITES BREVIFORMIS</small>, *var. γ*.

  l. Lateral view, and *d*, dorsal view, of a specimen from Glastonbury.
  s. Longitudinal section of a specimen from Eston Nab, in Yorkshire.

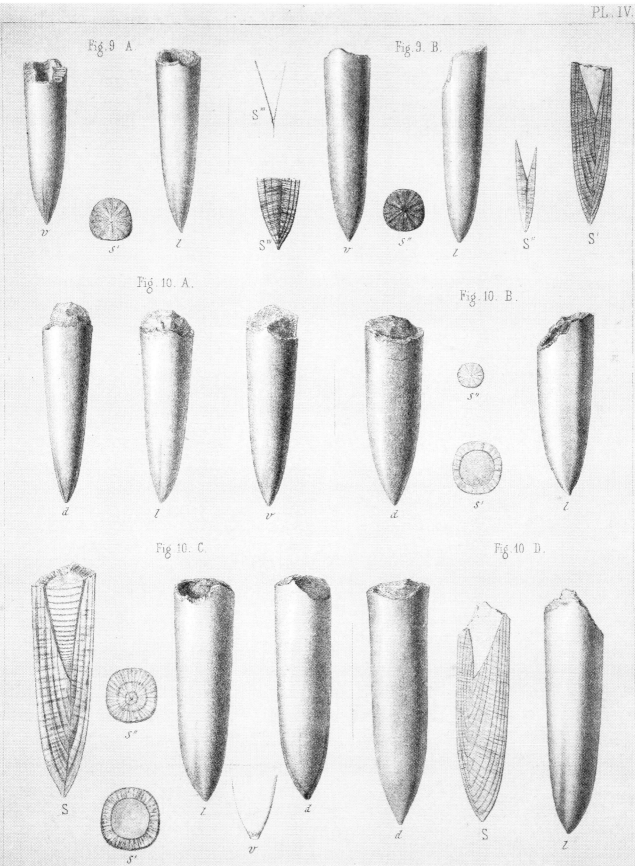

Fig. 9. A.

Fig. 9. B.

Fig. 10. A.

Fig. 10. B.

Fig. 10. C.

Fig. 10. D.

P. Lackerbauer lith.

Imp. Becquet, Paris.

EXPLANATION OF PLATE V.

FIG.

11. BELEMNITES GINGENSIS (Oolite).

   v. Ventral aspect of a specimen from the lower part of the Inferior Oolite of Dundry.
   l. Lateral aspect.
   s. Longitudinal section.
   f. Flanges of the phragmocone.
   Σ′. Siphuncle crossing the septa.
   Σ″. Same, enlarged.

12. BELEMNITES INSCULPTUS (Oolite).

   l′, l″. Lateral views of two specimens from the lower part of the Inferior Oolite of Dundry.
   v. The ventral aspect.
   s. Longitudinal section.
   s′. Transverse section across the alveolar cavity.

13. BELEMNITE from the Lias, allied to *B. insculptus*.
   d. Dorsal aspect of a specimen from the Lias of Lyme Regis.
   l′. Lateral view of the same.
   l″. Lateral view of a different specimen, with a very short axis to the guard.
   a. View of the end, to show the two distinct grooves.

14. BELEMNITES (LATISULCATUS.)
   l. Lateral view.
   v. Ventral aspect.
   s′. Section across the guard, through the alveolar cavity.

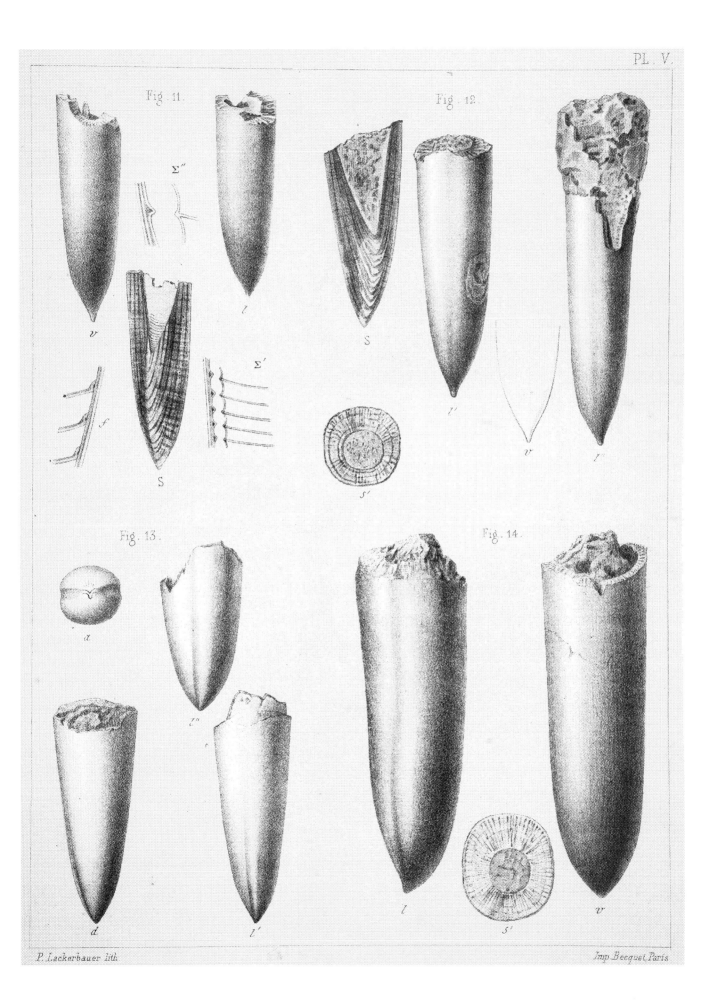

Fig. 11.

Fig. 12.

Fig. 13.

Fig. 14.

P. Lackerbauer lith.

Imp. Becquet, Paris.

EXPLANATION OF PLATE VI.

Fig.

15. BELEMNITES PAXILLOSUS.

l.   Lateral view,

v.   Ventral aspect, } of specimen with eroded and striated apex.

d.   Dorsal aspect,

s.   Longitudinal natural section of the guard, with a portion of the phragmocone not in section.   Natural sections are frequent in this species.

s′.   Longitudinal section of a younger individual.

s″.   Transverse section of the guard at the apex of the alveolus.

s‴.   Transverse section of the guard near the end.

φ.   Phragmocone, seen laterally.

16. BELEMNITES APICICURVATUS.

l.   Lateral view, showing the short groove near the point.

v   Ventral aspect.

d.   Dorsal aspect.

s.   Longitudinal section.

s.   Section across the alveolar region.

s‴, s^{iv}, s^{v}. Sections nearer and nearer to the end.

φ.   Side view of the phragmocone, always more arched than in *B. paxillosus.*

σ.   Striæ magnified.

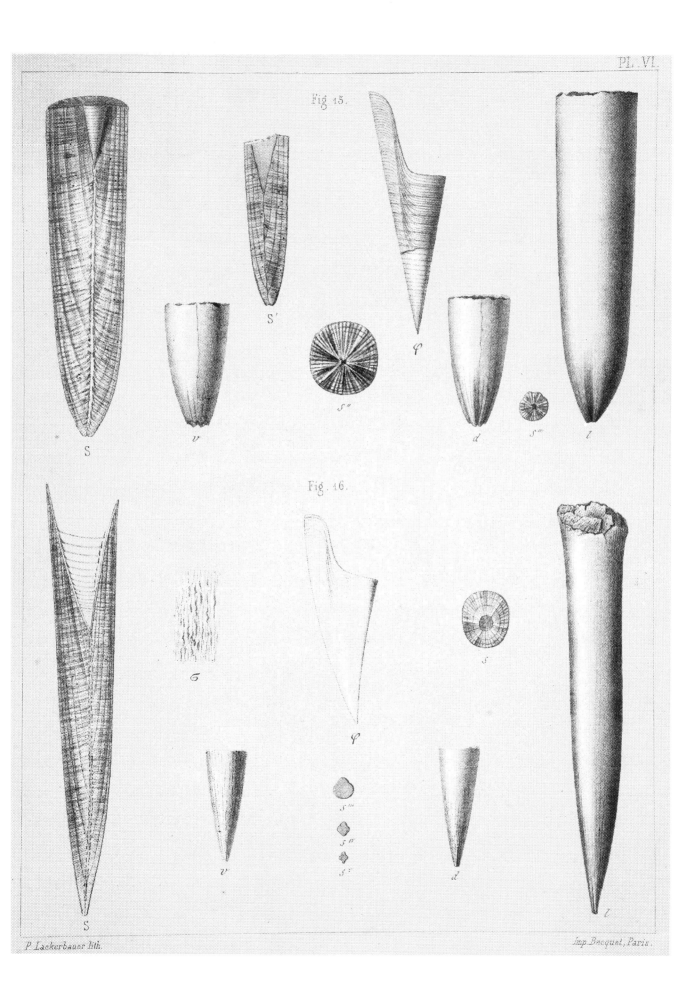

PL. VI.

Fig. 15.

Fig. 16.

P. Lackerbauer lith.

Imp. Becquet, Paris.

EXPLANATION OF PLATE VII.

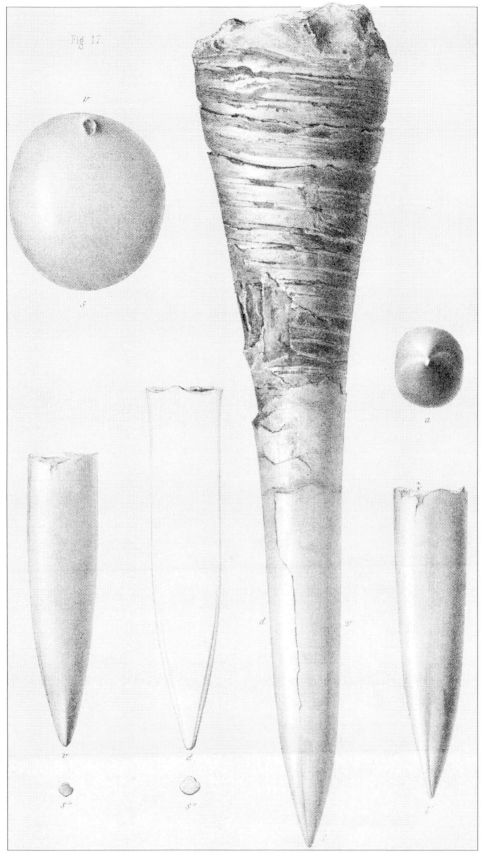

Fig. 12.

A MONOGRAPH

OF

BRITISH BELEMNITIDÆ.

BY

JOHN PHILLIPS,

M.A. OXON., LL.D. DUBLIN, LL.D. CAMBRIDGE, D.C.L. OXON., F.R.S., F.G.S., ETC.,
PROFESSOR OF GEOLOGY IN THE UNIVERSITY OF OXFORD.

PART III,

CONTAINING

PAGES 53—88; PLATES VIII—XX.

LONDON:

PRINTED FOR THE PALÆONTOGRAPHICAL SOCIETY.

1867.

PRINTED BY J. E. ADLARD, BARTHOLOMEW CLOSE.

BELEMNITES BUCKLANDI, n. s.   Pl. VIII, fig. 18.

Reference.   *Belemnites ovalis*, Buckland, 'Bridgew. Treatise,' vol. ii, p. 69, pl. xliv′, figs. 7, 8, 1836; and 'Bridgew. Treatise,' vol. ii, p. 71, pl. lxi, figs. 7, 8, 1858.

GUARD.   Slender, smooth, subhastate by reason of a gentle swelling toward the end, which is convexo-conical and somewhat obtuse, the sides meeting at an angle of about 45°, with scarcely a trace of striæ or grooves.

Transverse sections are nearly circular, and show the axial line nearer to the ventral side, which in some specimens is slightly but distinctly flattened.

Greatest length observed 5·15 inches, of which the axis of the guard occupies 2·00; the chambered part of the phragmocone appears to occupy the remaining portion of the length (3·15 inches). The greatest diameter below the apex of the alveolar cavity 0·50 inch.

Proportions.   The ventro-dorsal diameter at the alveolar apex being 100, the transverse diameter is 96, the ventral radius about 43, the dorsal radius about 57; axis, 480 in young, 300 in old.

PHRAGMOCONE.   Insufficiently known from Dr. Buckland's figure, which represents it with nearly straight sides meeting at an angle of 28°, and septal intervals about ⅟₇th of their diameter—an ordinary proportion.

Localities.   Golden Cap, Lyme Regis, from the Belemnite-bed (*Miss Anning, Phillips*). The first specimen figured by Dr. Buckland was in the collection of Miss Philpotts. In Upper Lias (sandy part), Blue Wick (*Phillips*). It has not occurred to me in the Midland Lias, nor do I find it in the large series of Belemnites from the vicinity of Banbury belonging to Mr. Stuttard.

This species has no slight analogy to *Belemnites ventro-planus* of Voltz ('Obs. sur les Bélem.,' pl. i, fig. 10), and to *B. umbilicatus* of Blainville ('Mém. sur les Bélem.,' pl. xi, fig. 11). Possibly they are all one species; for the umbilication of the end and the ventral flattening toward the end are not constant characters, and are not absent from some English specimens of *B. Bucklandi* which I have examined.

Observations.   The opinion of Dr. Buckland, that the extinct Belemnites and the living Sepiadæ agreed in possessing an ink-bag, was corroborated by the discovery of the specimen figured in his 'Bridgewater Treatise,' and copied fig. 18, Pl. VIII of this Essay. The collection of Miss Philpotts contained this, at that time, "unique specimen," which presents, in a somewhat indistinct roundish mass, what remains of the ink-bag in the anterior part of the phragmocone. "A fracture at *b′* shows the chambered areolæ of the alveolus. At *e* the thin, conical, anterior, horny sheath originates in the edge of the calcareous sheath, and extends to *e″*. The surface of this anterior sheath exhibits wavy transverse lines of growth; it is much decomposed, slightly nacreous, and flattened by pressure. Within this anterior conical sheath the ink-bag is seen at *e*, somewhat decomposed, and partially altered to a dark gray colour." ('Bridgewater Treatise,' vol. ii, p. 69, Ed. 1; and vol. ii, p. 71, Ed. 2.) The treatise just referred to contains several good representations of ink-bags, supposed to be of the "Belemno-sepia," which may have belonged to this species. The fine, almost complete, fossil animal figured by Prof. Huxley ('Mem. Geol. Surv.,' Monogr. II, pl. i, fig. 1) is not dissimilar to the specimen in Miss Philpotts' Collection, in all the sheath and phragmocone, but it is believed to belong to a different species.

The name assigned to this species in the 'Bridgewater Treatise' by Dr. Buckland, having been already employed by Blainville for a very different form, cannot be retained. Neither "*B. umbilicatus*" nor "*B. ventro-planus*" appears very suitable or of determinate application, so that, perhaps, the exigencies of the case may be best met by a new name, in honour of a great and early palæontologist.

BELEMNITES MILLERI, n. s. Pl. VIII, fig. 19.

GUARD. Slender, elongate, cylindroidal below the alveolar region, evenly tapering to a convexo-conical or acute summit, with none or only very faint traces of striæ or dorso-lateral grooves; section nearly circular, with the axis more or less excentric.

VAR. *a*. Apex convexo-conical, without trace of grooves.

β. Apex more pointed, with traces of very short dorso-lateral grooves.

Greatest length observed, 4·5 inches, of which the axis of the guard is 2·75.

Proportion of axis to diameter at apex of phragmocone 450 to 100 in var. *a*; 750 to 100 in var. *β*.

PHRAGMOCONE. Only known by a longitudinal section, which exhibits septa more than usually approximate, and sides somewhat arched, uniting at an angle of 28°.

Localities. Golden Cap, near Lyme Regis, in Middle Lias; Blue Wick, Yorkshire, in Upper Lias (*Phillips*). Hatch, near Taunton, in upper part of Lower Lias, with *Ammonites obtusus*, *A. raricostatus*, and *Spirifer Walcotii* (*Moore*). Lower Lias shales, the Belemnite-bed, Cheltenham (*Buckman*).

Observations. To judge by the drawing given by Miller to represent his *Belemnites elongatus*, this might have been the species meant, without grooves or striæ. Mr. Sowerby's suggestion of the relationship of *B. clavatus* (*pistilliformis*, 'Min. Conch.') to *B. elongatus* might thus acquire more probability, but the sections which I have made do not show in the inner laminæ the very clavate form of the young which is required by the hypothesis. In Quenstedt's pl. xxix, fig. 51, we have a slightly hastate form much allied to this, the specimen having been derived from the "black Alpine limestone," locally associated with talc-schists, or coal-formation, and Upper Liassic shales, near

Grenoble. Similar, also, are some Belemnites discovered by Mr. C. Moore at Camerton, in the "Bucklandian beds" of Lower Lias, of which it seems proper to add the provisional description which follows; though, until further specimens come to hand, sufficient figures and sections cannot be given.

BELEMNITES GRANDÆVUS, n. s.

GUARD. Slender, cylindrical, gradually tapering to a produced (bent) point; no groove on any part, but the sides somewhat flattened with age; section nearly round in the young specimen, a little compressed when older.

Dimensions. Of the two specimens, one, the younger, is 3·125 inches long; the other, older, only 2·5, probably a deformed specimen.

Proportions. The longest diameter at the apex of the phragmocone being taken at 100, the ventral part of it is 43, the dorsal 57; the apicial line 500 in the young specimen, but in the old specimen 300.

Locality. Lower Lias, "Bucklandian beds," Camerton, Somerset (*Moore*). The specimens are reddened by oxidation of iron.

Observations. Both specimens are bent towards the point a little irregularly, that is, obliquely to the general dorso-ventral plane of symmetry. The younger specimen has a produced striated apex; in the older one this part is contracted and irregularly plicated, with a sort of umbilicus. The surface is in parts eroded, so as to show the curiously undulated plications of the formative membrane. In early age the guard was depressed, when full grown compressed; these variations depend on the thicknesses of the successive increments by laminæ of growth.

BELEMNITES POLLEX, *Simpson.* Pl. IX, fig. 20.

Reference. *Belemnites pollex*, Simpson, Yorkshire Lias, No. 18, p. 27, 1855.

GUARD. "Subcylindrical, short, one side rather flattened; apex very obtuse, with a short irregular groove." "Length 3½ inches, width 1¼ inch."

Locality. Whitby, in Upper Lias, from the collection of the late Mr. Ripley (*Simpson*). West of Staithes, from the upper part of the Lower Lias (*Phillips*).

Observations. Of this remarkable form only two examples of the guard are known to me—the large specimen in the Whitby Museum, represented in my sketch, Pl. IX, fig. 20, which was described by Simpson; and the smaller one figured on the same plate, which is in my cabinet. Simpson's description is given above. The short irregular groove mentioned cannot be esteemed diagnostic. The diameter of the guard is almost imperceptibly enlarged behind the alveolar region. The axis of the guard does not much exceed the diameter at the alveolar apex. I hope to receive more specimens, and to give sections hereafter.

BELEMNITES ACUMINATUS, *Simpson.* Pl. IX, figs. 21, 22.

Reference. B. *acuminatus*, Simpson, Lias Fossils, No. 29, p. 29, 1855.
B. *ferreus*, Simpson, Lias Fossils, No. 28, p. 29, 1855.

GUARD. "Cylindrical for nearly the whole length, then suddenly ends in a sharp point somewhat produced; transverse section circular; length of guard about five times the diameter." To this description of B. *acuminatus* Simpson adds the following note, which, perhaps, refers to a different species:—"A specimen like in form has two slight grooves and one strong groove at the apex." Of B. *ferreus* he only says, "Apex elongated; no groove; base widened."

Locality. B. *acuminatus*, Jet-rock, Upper Lias, Whitby (*Simpson*).
B. *ferreus*, Middle Lias (*Simpson*).

The specimens are in the Whitby Museum.

Observations. The sketches given in Pl. IX of the two forms here described show their great affinity and probable identity. I have had no opportunity of examining sections.

On account of general form and freedom from grooves or striæ, there seems reason to maintain for these Whitby specimens a distinct place and name. I have not yet perceived among the fossils of Lyme Regis, Cheltenham, or Banbury, any closely allied forms; nor did the rich collections of Strasburg and Paris suggest to me any decided analogue.

ON A GROUP OF ELONGATE BELEMNITES, WITH STRIATED APICES.

Among the frequent fossils in the scars at Saltwick, near Whitby, where the Upper Lias shales are largely exposed, is a beautiful group of slender elongated Belemnites. For the most part they exhibit but feeble traces of lateral or ventral furrows, but are more than usually striated at and near the apex. These forms are represented on Pl. X, which includes several species or varieties, as may be determined by further research. Figs. 23, 24, 25, 26, show forms in which the apicial part is conoidal and deprived of distinct furrows; while in Fig. 27 the usual furrows appear, and the Belemnite is tripartite. All of them are acute; some have the apex very much drawn out; all are striated near the apex, and for some distance from it. All are somewhat compressed, some of them considerably. In the published figures of Belemnites none are so like to these as the representations of *Belemnites acuarius gracilis*, Quenst., 'Céph.,' pl. xxv, fig. 4, and *Belemnites tripartitus gracilis*, by the same author, pl. xxvi, fig. 17.

The fossils so depicted come from the Upper Lias; they have distinct ventral and dorso-lateral grooves.

Several of these Yorkshire forms have been described by Simpson (Lias Fossils, No. 9, 10, 11, 12, 16), but not figured.

In the cliffs of Dorsetshire, about Lyme Regis, Charmouth, and Seatown, a considerable number of Belemnites occur corresponding to these in length, general form, and compression; some are deficient of grooves near the apex, and others show them more or less; but striation about the apices, if not wholly absent, is not systematically present. The surface of the Dorsetshire Belemnites is not always perfectly preserved; it is often somewhat eroded. I have, however, succeeded in identifying a few specimens with one of the Yorkshire fossils. Besides this, the sides are usually channelled or marked by plane facets, so as to produce on the whole a different aspect to the eye. Some of these forms from the Dorsetshire Lias are given on Pl. XIII.

After examining as many specimens as I could extract from the Lias near Whitby, and others from Robin Hood's Bay, collected for me by Mr. Peter Cullen, I have thought it desirable to employ separate names for four forms which offer available characters, and perhaps further search may prove them to be really distinct; but it will be often very useful to employ for them all collectively the general term of "*graciles.*" There is one allied form in the Lower Lias of Yorkshire.

BELEMNITES LÆVIS, *Simpson.* Pl. X, figs. 23, 26.

Reference. Belemnites lævis, Simpson, Lias Fossils, No. 9, p. 25, 1855.
Belemnites trivialis, Simpson, Lias Fossils, No. 11, p. 26.

GUARD. Compressed; elongate; tapering uniformly, and at last rather suddenly, to the apex, which is occasionally crested, plaited, or striated, without distinct grooves on any face. Sections show the contour to be elliptical, with a slightly excentric, nearly straight axis.

Dimensions. Greatest length observed of the guard alone 4 inches, beyond which the conotheca, converted into bisulphide of iron, is traced 1 inch.

Proportions. The diameter, $v\,d$, at the apex of the phragmocone, being taken at 100, the cross diameter is 88, the ventral radius 44, the dorsal 56, the axis 350 to 500.

PHRAGMOCONE. A little arched, ending in a spherule, with septa occupying a length equal to that of the axis of the guard.

Locality. Jet-rock of the Upper Lias, Whitby, rather plentiful (*Simpson*); also above the Jet-rock at Whitby and Robin Hood's Bay (*Phillips*). Specimens allied to these occurred to me in Lower Lias (upper part) under Huntcliff, but they are very much more acute (see *B. elegans*).

Observations. Simpson makes no mention of the phragmocone or alveolus, but his notes on the guard are ample. They are as follows:

" *B. lævis.* Length of guard about five times its width, slightly depressed, smooth, and regular; apex generally compressed, sometimes with a flattened process; the surface in places toward the smaller end firmly and distinctly corrugated like leather, probably by the impression of the mantle; general length 3 or 4 inches. Some have the apex striated."

" *B. trivialis.* Subconical, moderately stout, expanding at the base, with numerous fine striæ at the rather obtuse apex; general length 2 or 3 inches; smaller specimens approach the slender *B. subtenuis*, larger ones the more robust *B. lævis*."

The figures given in Pl. X, fig. 23, were sketched from specimens in the Whitby Museum; the others (fig. 26) are in my cabinet, collected from the Saltwick shore (1864). The striation is remarkable for distinctness and continuity over about a third of the post-alveolar space; but it can be traced further forward in short, discontinuous, wavy ridges.

BELEMNITES DORSALIS, n. s.   Pl. X, fig. 24.

GUARD.   Cylindroidal or slightly subhastate, tapering very gradually to a sharp apex; lateral grooves or flattenings on the middle and alveolar regions, not reaching to the apex; no ventral groove; a short definite dorsal groove on the alveolar region only.

Sections show the contour to be nearly circular, or a little compressed, or a little depressed; the axis excentric, even remarkably so in some examples. Greatest length, including expansion of the guard, $2\frac{3}{4}$ inches; greatest diameter behind the alveolar cavity, $\frac{1}{4}$ inch.

Proportions. The diameter, $v\,d$, at the alveolar apex being taken at 100, the radius v is from 30 to 40 (on an average about 35), and the radius d 60 to 70 (on an average 65); the axis about 700.

PHRAGMOCONE.   Insufficiently seen for accurate description.

Locality. Saltwick, near Whitby, in Upper Lias (*Phillips*). Filey Cliffs, in drifted Lias (*Phillips*).

Observations. Thirteen specimens of different magnitudes, which I collected at Saltwick in 1864, furnish the grounds for the determination of this curious and rare form. I have seen it in no other collections, and have no knowledge of it beyond the small, probably young, specimens here figured. The little deep slit or canal on the dorsal aspect of the guard, is a circumstance unexampled in the Oolitic or Lias rocks. It appears only on the retral part of the alveolar cavity, and ends in most specimens quite distinctly and definitely, both forward and backward. It is ¼ inch long. The lateral grooves or flattenings are variable in distinctness.

BELEMNITES STRIOLATUS, n. s.   Pl. X, fig. 25.

GUARD. Compressed, elongate, tapering gradually to an attenuated, conical, finely striated apex; the striæ extend continuously over half the post-alveolar space, and in well-preserved specimens can be traced onward into minute, interrupted, undulated striæ, over nearly the whole surface. Where best seen (one third of the post alveolar space from the apex), there are about 100 striæ.

Sections across the guard behind the alveolus are elliptical, the axis but slightly excentric.

Dimensions. The largest observed is 3½ inches long; of this length ¾ inch is crushed over the alveolus; greatest diameter (uncrushed) less than ½ inch. The smallest specimen is less than 2 inches in length.

Proportions. Taking the diameter from back to front at the alveolar apex at 100, the cross diameter is 85, the ventral radius 45 or more, and the dorsal 55 or less; the axis is therefore but little excentric, and measures 500.

Locality. On the Upper Lias Scars at Saltwick, near Whitby, and in the same beds at Robin Hood's Bay (*Phillips*). A small number of specimens from the Belemnite-beds under Golden Cap are identified with this species (*Phillips*). A few small young examples, Glastonbury (*Moore*).

Observations. The affinity of this to the last-described species is obvious; and it is quite possible that further research may unite several of the forms here named separately into one species. But that will not render useless the nomenclature, if it expresses real and often observable peculiarities, for these are elements in the problem of the variation of life-form, in relation to time, space, and physical condition.

It is doubtful whether this species has been described. Simpson's description of *Belemnites substriatus* agrees, indeed; but the fossil is much larger than my specimens, an example in the Whitby Museum being 5⅓ inches long.

"Similar in general form to *Bel. trisulcosus*, but with numerous striæ at the apex, and no grooves." (Lias Belemnites, No. 16.)

Belemnites subtenuis of Simpson contains, beside the trisulcose form which is characteristic, two others which are described as varieties, viz.—

Var. *a.* Grooves obsolete, apex not striated.
b. Thicker in proportion, apex not striated.

This want of striation is the only obstacle which seems to forbid the otherwise probable union of the varieties *a* and *b* to *B. striolatus.*

BELEMNITES SUBTENUIS, *Simpson.*   Pl. X, fig. 27.

Reference.   Belemnites subtenuis, Simpson, Lias Fossils, No. 12 (excl. *a* and *β*), p. 26, 1855.

GUARD.   Very elongate, compressed, tapering uniformly to the apex, which is marked by three faint, though rather long furrows, and many fine striæ.

Sections show the outline to be elliptical, the axis very excentric and straight.

Dimensions.   Largest specimen in my collection 4 inches long, the greatest diameter (uncrushed) less than $\frac{1}{2}$ an inch.

Proportions.   The diameter, *v d,* at the apex of the phragmocone being taken at 100, the ventral radius is 36—38, the dorsal 62—64, the axis 700 to 1000, the diameter from side to side 76.   Nearer the apex the diameters are less unequal.

PHRAGMOCONE.   Visible, but not distinctly observable in the expanding anterior region. The flanges of the septa appear very short; the cross section is elliptical, but not so much as the section of the guard, which is thinner on the sides than on the dorsal or ventral face.

Locality.   Above the Jet rock in Upper Lias at Whitby, abundant (*Simpson, Phillips*). Robin Hood's Bay, in Upper Lias (*Cullen*).

Observations.   Simpson's description of *B. subtenuis* is in the following words :— " Slender, regularly tapering, with three long shallow grooves toward the finely striated apex."   He allows two varieties from the type, viz. :—

Var. *a.* Grooves obsolete, apex not striated.
b. Thicker in proportion, apex not striated.

These may probably be better referred to *B. striolatus* (see p. 59).   The group thus becomes definite, and may be compared with Simpson's larger but similar straight tripartite forms, viz., *B. trisulcosus* and *B. incisus,* which, it appears possible, may be full-grown individuals of the same species (see also p. 62).

On Belemnites allied to Belemnites tripartitus of Schlotheim.

Under the name "*Belemnites tripartitus*" foreign palæontologists have assembled a considerable variety of forms, some distinction among them being made after the manner of Quenstedt, who, in his 'Cephalopoda,' employs as *general terms* such titles as *tripartitus, paxillosus, compressus, brevis, digitalis;* and qualifies them by addition of other terms, as in *B. tripartitus sulcosus*, or by joining together even general terms, as *digitalis tripartitus, tripartitus paxillosus*, and the like.   By this mode of proceeding the idea of real specific diversity is obscured, and that of a vague mixture of characters is introduced.   Yet it has some considerable advantages for a serious inductive study of a large Belemnitic series, and will be referred to again.   Others, as D'Orbigny, plainly join into one specific group a large number of "tripartite" forms which seem, at first view, to claim separation.   Thus, under *B. elongatus* this author enumerates *B. aduncatus*, Miller, *B. trisulcatus*, Hartmann, *B. oxyconus*, Hehl, *B. incurvatus*, Zieten, and *B. propinquus*, Münster. And in his plate viii, figs. 6—11, the species is called *B. tripartitus*, Schl., though that name is omitted in the text ('Terr. Jurass.,' p. 90).   Voltz gives as a different species *B. trifidus* ('Obs. sur les Bélem.,' pl. vii, fig. 3) ; and Blainville had already suggested for separation his *B. trisulcatus* ('Mém. sur les Bélem.,' pl. v, fig. 13).

On the Liassic coast of Dorsetshire these forms are so rare that I can hardly quote one from personal research.   One is referred to by Quenstedt as from Lyme Regis, with distinct ventral groove, under the title of *B. digitalis tripartitus*, and figured pl. xxvi, fig. 31, of his 'Cephalopoda.'   He remarks that it is exactly like German (*i. e.* Wurtemberg) examples.   The Yorkshire coast produces a greater number and a greater variety of such forms.

There appear to be two principal sections of them, which in mature age may be thus separated :

Elongate, straight, three-grooved, often striated, as *B. tripartitus* and *B. subtenuis* of these pages.

Shorter, somewhat recurved at the apex, with three rather short grooves, and few or no striæ, as *B. subaduncatus* and *B. vulgaris*, to be noticed hereafter.

In Mr. Simpson's work on the Lias of Yorkshire the "tripartite" forms are employed to constitute a larger number of species than appear to me necessary ; but though I have had the opportunity of inspecting the fine collection at Whitby, which is the basis of his work, I do not find in it a sufficient *series* of forms, from youth to age, of the supposed species, to give more than a few indications of the synonymy which may guide further research.

BELEMNITES TRIPARTITUS, *Schlotheim*.   Pl. XI, fig. 28.

> *Reference.*   *Belemnites tripartitus*, Schlotheim, 'Petref. Belem.,' No. 6, p. 48, 1820.
>
> > *B. elongatus*, D'Orbigny, 'Pal. Franç. Terr. Jur.,' p. 90, pl. viii, fig. 11, 1842.
> >
> > *B. digitalis tripartitus*, Quenst., 'Ceph.,' p. 419, pl. xxvi, figs. 14, 31, 1849.
> >
> > *B. trisulcosus*, Simpson, 'Lias Fossils,' No. 14, p. 26, 1855.   (Section ovate.)
> >
> > *B. incisus*, Simpson, 'Lias Fossils,' No. 15, p. 27, 1855.   (Section nearly circular.)

GUARD.   Straight, elongate, cylindroidal; sides more or less flattened, tapering in a continuous curve in the post-alveolar region to a three-grooved apex, which is distinctly striated on the dorsal aspect for half the length of the axis of the guard; grooves deep-ening toward the apex.

Sections show the contour to be oval, with the sides rather flattened, the dorsal part rather broadest, except toward the apex, where the contrary happens; the axis excentric and straight.

Greatest length observed (behind the expansion of the alveolar cavity), $6\frac{1}{2}$ inches; greatest total length, 9 inches; greatest diameter, 1 inch.

Proportions.   Taking the diameter, $v\ d$, at the apex of the phragmocone at 100, the ventral radius is 37, the dorsal 63, the cross diameter 88, and the axis 550 to 600.

PHRAGMOCONE.   Not seen in my specimens.   The section of the alveolus is very slightly oval.

Locality.   In the Upper Lias of Saltwick (*Phillips*).   In the Jet-rock, Saltwick (*Simpson*).   In the Middle Lias of Banbury (*Stuttard*, No. 87).

Observations.   In its young state this species is not certainly known, unless, as I think not improbable, *B. subtenuis* holds that place.   Its very distinct, straight, narrow grooves, continued to the apex, distinguish it from the other triglyphic Belemnites of the Upper Lias of Yorkshire.   Its nearest relative is a beautiful species which occurs in the Upper Lias of Ilminster and other localities of the South of England.   Neither of those species has been found (as far as I have seen) at Lyme Regis.   A specimen in my possession shows, on the dorsal aspect near the apex, a short, very narrow groove, which does not reach the apex (see Pl. XI, figs. 28 *d* and *s″*).   In a younger specimen the striæ are interrupted, as in fig. 28 *σ*.

BELEMNITES SUBADUNCATUS, *Voltz.* Pl. XI, fig. 29.

Reference. *Belemnites subaduncatus,* Voltz, 'Obs. sur les Bélemn.,' p. 48, pl. iii, fig. 2, 1830.
       B. expansus, Simpson, ' Lias Belemnites,' No. 39, p. 46, 1855.

GUARD (Young). Cylindroidal, or subprismatic in the alveolar region, thence tapering in a lanceolate form (so as to be slightly or even distinctly subhastate) by a gentle curve to a very pointed prominent end, which bends a little toward the dorsal side. Three grooves part from near the ungrooved end, two of them dorso-lateral, becoming distinct at a short distance from the end, and gradually vanishing before reaching the alveolar apex ; the third medio-ventral of variable length and distinctness, usually short, but perhaps never quite absent.[1]

(Old.) Few examples are certainly known ; in them the figure is more cylindroidal and more compressed, the termination is less acute and more recurved, so as to resemble, except in greater length, mature individuals of *B. vulgaris,* from which in youth they are quite distinct.

Longitudinal sections show the apicial line to be somewhat curved, nearest the ventral face, and three, four, or five times as long as the diameter, according to age. Transverse sections in the young are subquadrangular across the alveolar region, somewhat oval behind it, and trilobed near the end ; in the older specimens this section is decidedly oval. In all examples the axis is excentric, in the young remarkably so ; the dorsal laminæ usually thickest over the alveolus, but the contrary also occurs.

Dimensions. The smallest specimen in my collection is 2 inches long (of which the axis is $1\frac{1}{2}$), with a diameter of $\frac{1}{4}$. The sizes are traced with certainty to a diameter of $\frac{3}{4}$ inch, with an axis of $1\frac{3}{4}$. Other examples have a longer axis; but, on the whole, it appears, the proportions grow more robust with age. Voltz describes and figures specimens of intermediate magnitude only. I am uncertain as to the really old forms of the species; but I believe *B. distortus* of Simpson, No. 31 (Whitby Museum), to be a good example. On the whole, we may be sure of the identification of the young and middle-aged forms so common on the Whitby Scars, and so remarkable for their slender shape, very acutely pointed apex, slightly hastate figure (though this varies, and is sometimes only just traceable) by reason of a gentle swelling at about two thirds of the distance from the end toward the alveolar apex. The rather prismatic shape of the alveolar region caused Voltz to call it tetragonal, and sometimes the expression is correct.

---

[1] In my oldest specimen, $\frac{3}{4}$ inch in diameter, it is hardly traceable.

Proportions. Taking the diameter from back to front at the alveolar apex at 100, the cross diameter is about 90, but this varies in different specimens; the excentricity of the axis is variable, in some young specimens the ventral is only 28 to the dorsal 72; in another specimen the ventral is 40, the dorsal 60; in another, ventral 48, dorsal 52; the axis is from 300 in middle-aged to 500 or 600 in the young. In Voltz's figure the axis is 340.

PHRAGMOCONE. Very slightly arched; its section oval, with proportions of 100 to 93; septa close, rather unequally arranged, but in the middle part, on the average, distant one seventh of the diameter, apparently formed of a single plate with a short flange. In a specimen from Whitby an extraordinary number of the septa are squeezed together in the hinder part of the last chamber, by pressure from without. If these were supposed to be the last septa, and to be replaced, the axis of the phragmocone would equal half the axis of the guard. The angle appears to be 24°30′ (Voltz gives 25°).

Locality. In the Upper Lias shale of Whitby and Saltwick (*Phillips*).

BELEMNITES ILMINSTRENSIS, n. s.   Pl. XII, fig. 30.

GUARD. Elongate, straight, conoidal, more or less compressed, very gradually tapering over the whole of the guard to an acute point (24°), often cross-banded with light and dark shades; axis subcentral when young, more excentric when old; one ventral, two dorso-lateral furrows, all reaching the apex, the ventral furrow usually longest of the three. In young specimens the guard is depressed about the end, and shows a broad ventral furrow there, with slight and short traces of the dorso-laterals. The sides are often slightly canaliculate for the whole length in young specimens.

Dorsal aspect rather broader than the ventral.

Sections show great excentricity of the axis in full-sized specimens, and in adults an oval section. In young specimens which are depressed, with flattened sides, the section of the guard is subquadrate.

Dimensions. Greatest length of axis of guard observed, 5 inches; greatest diameter of the most expanded part, 1 inch.

Proportions. Taking the dorso-ventral diameter at the alveolar apex at 100, the ventral radius is 33 +, the dorsal 66 +, the cross diameter 85; the axis in a very long compressed variety (axis 4·5 inches long) 800, in a shorter variety 600, in one still shorter 400, in the shortest 350. These are adults. In a young specimen the axis (1 inch long) is 350, in another (axis 0·6 inch long) 300, in the shortest of all the specimens (less than ½ inch long) the proportion is nearly the same.

PHRAGMOCONE. Much extended under an angle at first of 24°, and afterwards of 18°.[1] Septa formed of one lamina, elliptical, with axis as 100 to 110, numerous (60 or more); depth of chamber, $\frac{15}{100}$ of the diameter; inner lateral surface of the chamber smooth and flat below the septa. Conotheca slightly undulated in rings, a little concave opposite each septum, a little convex between the septa. The septal outline is waved, and descends to the siphuncular border.

Greatest diameter observed, 1·6 inch; greatest length, $4\frac{1}{4}$ inches; in this case the axis of the guard is about $2\frac{1}{8}$ inch, and there are about 50 septa.

Locality. Ilminster, in Upper Lias, abundant, and of all ages (*Moore*). Dundry, full-grown, Upper Lias (*Bristol Museum*). Glastonbury (*Phillips*). Kimberley's Brickyard and Workhouse Yard, Banbury, in Upper Lias, with *Ammonites communis* (*Stuttard*, No. 42, 44, 45). Upper Lias, Stroud (*Buckman*).

Observations. This elegant species is remarkable for the continual tapering through its whole length, by which it happens that no part of the guard is really cylindroidal. The angle of inclination of the sides of the phragmocone is very moderate (not exceeding 18° in the anterior part, but amounting to 24° in the hinder part); the section is elliptical.

In Mr. Moore's rich collection from Ilminster the growth of this species may be traced with great satisfaction from an individual less than half an inch long to full-grown examples of 6 and 8 inches from the apex of the guard to the last (or nearly the last) chamber. Two varieties also appear of unequal proportions, one being much more compressed, and with a longer guard. This is less common than the shorter variety. The degree of distinctness of the furrows also varies, so that we may mark different races; the longer ones, indeed, may be (according to D'Orbigny) males, the shorter ones females.

Var. *a.* All the three furrows of the guard distinct, the ventral one usually longest.

β. All the furrows indistinct.

And to each of these the variations of length may be added. Striations can hardly be traced about the apex, but occasionally appear in the ventral sulcus.

Taken as a whole, it appears that this fossil more than any other resembles in shape and proportions the original figure given by Miller for *B. elongatus;* but no specimen corresponding to that figure has been found in the Bristol collection. On the other hand, there are in that collection thick, tripartite Belemnites, with short axis of guard, referred to *B. elongatus* of Sowerby, from the Upper Lias at Dundry, with an alveolar angle of 25°.

---

[1] In one case this angle in the anterior part of the cone is found to be only 12°30'; in another large specimen, as much as 24°; both exceptional instances.

BELEMNITES MICROSTYLUS, n. s.   Pl. XIII, fig. 31.

GUARD.   Very slender, almost perfectly cylindrical in the post-alveolar region till towards the apex, expanding over the conotheca with remarkable regularity.   No lateral groove.

In young specimens the substance of the guard is transparent and solid toward the point, mostly opaque, and more friable in a long space over and behind the alveolar region, as in *B. clavatus.*

PHRAGMOCONE.   Straight, very regularly tapering at an angle of 18°, with septa even or scarcely waved, and placed at distances somewhat greater than usual, viz., one fifth of the diameter.

Section of the phragmocone slightly elliptical.

Locality.   One specimen, No. 349, in the Collection of the Geological Survey, Jermyn Street (fig. 31, G), was discovered by Mr. Day in a nodule from the Belemnite-bed of Lyme Regis.   Another in the Oxford Museum (fig. 31, o) was presented by Mr. Murley from the Insect-bed at Dumbleton.

BELEMNITES LONGISSIMUS, *Miller.*   Pl. XIII, fig. 32.

Reference.   *Belemnites longissimus,* Miller, 'Geol. Trans.,' 2nd series, vol. ii, p. 60, pl. viii, figs. 1, 2 (Paper read April, 1823), 1826.

GUARD.   Excessively elongate, compressed, with faint lateral grooves, and a blunt, roughly striated apex.

PHRAGMOCONE.   Unknown.

Locality.   Lyme Regis.   In the Collection of the Bristol Institution is a specimen (No. 33) which may have served Miller in considering the species named *B. longissimus.* Another in the same collection (No. 27), marked *B. cylindricus,* Blainville, may be identical.   I have a few examples from the Belemnite-bed under Golden Cliff, but none of these specimens are sufficient for good description.   A Belemnite of the same general form, and even longer in proportion when young, occurs in shales of the Cromartie Coast, which have been called " Lias " by geologists.   They, however, contain only fossils of the Oxford Clay and Coralline Oolite series, and the Belemnite has a different shape when old.

BELEMNITES JUNCEUS, n. s. Pl. XIII, fig. 33.

GUARD. Excessively elongated, slender, nearly of equal diameter for a great length, expanded anteriorly, compressed, with shallow lateral risings and hollows extending along the whole length of the surface.

Sections of the sheath show the sparry lamellæ nearly circular toward the axis, but very elliptical toward the circumference. The apicial line is nearly central. The fibres are coloured in light and dark bands. (The summit is unknown; the compression increases toward the summit.)

PHRAGMOCONE. Unknown.

Dimensions. Total length of a fragment, above 3·5 inches; greatest diameter, 0·33; axis of guard, above 3·0.

Proportions of the guard, the diameter $a\,b$ at the apex of the alveolus being 100, the ventral radius is 48, dorsal radius b 52, cross diameter 75 to 85, axis of guard above 1500.

Locality. Golden Cliff, near Lyme Regis. The author possesses portions of two guards, including the alveolar and parts of the apicial regions.

BELEMNITES NITIDUS, n. s. Pl. XIII, fig. 34.

GUARD. Remarkably elongated, compressed, for a great portion of its length of equal diameter, but acutely conical toward the summit, which is pointed; expanded anteriorly; smooth, with two very slight dorso-lateral grooves near the summit, and a double lateral furrow extending over nearly all the apicial region.

Sections show the apicial line nearly central, straight, surrounded by subelliptical lamellæ, finely fibrous, the outer ones acquiring double lateral flutings.

PHRAGMOCONE. Unknown.

Dimensions. Total length, 7·5 inches (8·0); greatest diameter, 1·0; axis of guard, 5·5; alveolar axis, probably 2·5.

Proportions of the sheath, the diameter $a\,b$ at the apex of the alveolus being 100, the ventral radius is 46, dorsal radius 54, cross diameter 80, axis 1000.

Locality. Belemnite-bed, Golden Cliff, Lyme Regis (*Oxford Museum, Phillips's Cabinet*). Banbury, Lower Lias (*Stuttard*, No. 135).

BELEMNITES QUADRICANALICULATUS, *Quenstedt*. Pl. XIII, fig. 35.

Reference.   *Belemnites quadricanaliculatus*, Quenstedt's 'Jura,' p. 285, pl. xli, fig. 17,
       1857.

GUARD.   Long conical, marked by four narrow grooves, from the apex to and
extending over the alveolar region.   One of the grooves is dorsal, and in very good
specimens appears double (fig. 35 *d'*); it is deepest on the alveolar region; two are
dorso-lateral, one is ventral.   The apex is somewhat blunt, the grooves reaching to it.
The surface unusually rough.

PHRAGMOCONE.   Unknown.   Alveolar angle not measurable in either of the two
specimens placed before the author by the Geological Survey.

Dimensions.   Total length of specimen, 1·75 inch; greatest diameter, 0·33; axis of
guard, 1·00 inch.
Proportions.   The ventro-dorsal diameter at the alveolar apex being 100, the axis of
the guard is 500, the cross diameter 96.

Locality.   Upper Lias Sands at Chidcock (Museum of the Geologic al Survey in
Jermyn Street, London).   Upper Lias, Ilminster (*Moore*).

BELEMNITES TUBULARIS, *Young* and *Bird*. Pl. XIV, fig. 36.

Reference.   *Belemnites tubularis*, Young and Bird, 'Geol. of Yorkshire Coast,'
                         p. 259, pl. xiv, fig. 6, 1822; Phillips's 'Geology
                         of Yorkshire, vol. i, p. 163, pl. xii, fig. 20,
                         1829.
       B. acuarius, Morris, 'Catal.,' p. 300, 1854.

GUARD.   Very elongate, slender, expanding over the alveolar region, uniformly
tapering or cylindroidal in the post-alveolar region, greatly flattened in the apicial region,
ventral and dorso-lateral furrows on the flattened part.   Surface striated unevenly, especially on the alveolar region.   Proportion of the flattened part variable in different
specimens.   A slight hollow or flat space often runs down the sides of the guard.   Strong
parallel striæ near the apex.
       Transverse sections show an oval outline, with a nearly central axis in the young, a
nearly circular outline in the older.

Dimensions. Greatest length observed (including only the beginning of the phragmocone), 10 inches; greatest diameter in the post-alveolar region, less than $\frac{1}{2}$ inch.

Proportions. Taking the diameter from back to front at the alveolar apex as 100, the ventral radius is 48, the dorsal 52, the cross diameter from 80 to 90, the axis of the guard about 1400.

PHRAGMOCONE. Incompletely observed; probably straight, with an angle of about 18°.

Locality. In Upper Lias shale, above the Jet-bed at Saltwick, near Whitby (*Phillips*). Sandsend, near Whitby (*Phillips*), and Robin Hood's Bay (*Cullen*).

A specimen in the Bristol Museum (B 1, 42), is said to be from Gloucester. It closely resembles, however, Yorkshire specimens, and I have never observed the species in any locality south of Yorkshire.

Observations. By German writers this is usually supposed to be identical with some one of the many forms included under the title of *Belemnites acuarius*. Quenstedt figures, pl. xxv, figs. 2, 9, 10, specimens from the Upper Lias of Ohmden, which he terms *B. acuarius tubularis*; they are uncompressed, and show a distinct ventral groove, and an undulated alveolar border. D'Orbigny includes *B. tubularis* among the many synonyms of *B. acuarius* with *B. longissimus*, Mill., *gracilis*, Hehl., *lagenæformis*, Hartmann, *longisulcatus*, Voltz, *tenuis*, Münst., *semistriatus*, Münst., *gracilis*, Römer. He regards the guard as subject to a remarkable extension retrally, at a certain age, previous to which it does not differ, he says, from that of *B. irregularis*. The extension is stated to be most frequently hollow, so that the term "tubularis" would be really deserved, and the compression to a flat plate easily explained. When this is not the case, from the filling of the cavity with calcareous matter (not fibrous), it constitutes the so-called *Pseudobelus* of Blainville. Such is the view of D'Orbigny, who was ready to unite *B. irregularis* and *B. acuarius* in one species, the former being females, which preserved always their original obtuseness, and had no retral extension, the latter being males. This remarkable opinion he bases on an examination of the recent *Loligo subulata*, the males of which have a very long dorsal plate (osselet), the females a short one. I propose to consider this subject in a general discussion, embracing other species; but at present it appears only necessary to say, that there is no ground for admitting the specific affinity of *B. irregularis*, or any forms like it, with *B. tubularis* of the Yorkshire coast.

The length of the unflattened part of this Belemnitic guard is very unequal in different specimens of about the same total length—one example in my possession gives eight inches for this part, several three inches, one two and a half. The flattened part in one example exceeds six inches. A fine specimen obtained by the Rev. Dr. Plumptre, at Whitby, and now in the Oxford Museum, is thus measured: total length 12 inches,

10

apicial region, grooved and flattened, 1 inch; post-alveolar region, uniformly tapering, $7\frac{1}{2}$ inches; alveolar region of visible guard, 2 inches; projection of phragmocone beyond it, $1\frac{1}{2}$ inch. The phragmocone is extended to a total length of nearly $3\frac{1}{2}$ inches, all chambered, with a terminal breadth, completely flattened, of $1\frac{1}{4}$ inch. This gives an apparent angle to the phragmocone of 27°; but if reduced to a cone, about 18°. The diameter of the guard at the alveolar apex is about 0·375 inch; the length of the axis, twenty times as great. Nearly all the surface is striated, the striæ being everywhere undulated and interrupted, and largest and most conspicuous on the alveolar region.

Another fine specimen also presented by Dr. Plumptre to the Oxford Collection is $11\frac{1}{2}$ inches long; of this the posterior $7\frac{3}{8}$ths inches are compressed flat; 3 inches remain quite unaffected by pressure, and beyond this is 1 inch of crushed alveolar cavity. Diameter at the alveolar apex $\frac{4}{10}$ths of an inch. Three parallel grooves, which seem to be too regular to be the effect of mere crushing, appear on a great part of the flattened post-alveolar region; toward the point only one is traceable, and it is accompanied by several striæ. The grooves referred to do not extend over the solid post-alveolar and alveolar tracts; on which, however, facettes can be traced which seem to correspond with the grooves. From this it seems, on the whole, probable that the grooves are the effect of uniform pressure on the unequally resisting facetted surface of the hollow part of the guard.

The shells which commonly accompany this Belemnite are chiefly *Posidonia* (*Inoceramus dubius* of the 'Min. Conch.'); and the shale may well be called, as in Würtemburg, where *B. acuarius* occurs in it, the Posidonian shale.

BELEMNITES ACUARIUS, *Schlotheim.* (Diagram No. 22.)

Reference. Schlotheim, 'Petref.,' p. 46, No. 2, 1820; D'Orb., 'Terrains Jurassiques,' pl. vii, referred to in his text, p. 76, as pl. v; Quenst., 'Jura,' p. 258, pl. xxxvi, fig. 9, 1857.

GUARD. Elongate, compressed, striated on the sides in the retral part, and flattened, by compression, at the end.

Transverse section oval, with somewhat flattened sides; the ventral surface broadest, axis nearer to the ventral side.

Taking the diameter from back to front at 100; the dorsal radius is 66, the ventral 34, the cross diameter 73.

PHRAGMOCONE not seen in English specimens.

Locality. Cheltenham, in the Belemnite-bed of the Lower Lias (*Buckman*).

Observations. In the figure quoted from Quenstedt the lateral striation is not

DIAGRAM 22.

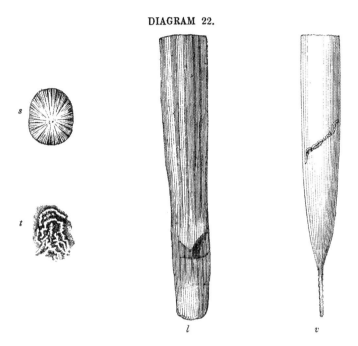

represented: in other figures of Quenstedt referred to *B. acuarius* ('Cephalopoda,' pl. xxv), lateral striation is represented, but not the terminal compression. D'Orbigny, pl. vii, fig. 4, represents the retral part of the guard as tubular. These fossils belong to the Upper Lias. At *t* the winding interior structure is represented.

On Belemnites allied to *B. DIGITALIS* of Schlotheim, Plates XV, XVI.

Few Belemnites appear on a first view more characteristically separate from the rest than the compressed, very blunt, straight-sided forms of the Upper Lias, which are frequent in the continental museums, under the name of *B. digitalis*. In only a few instances is this form exactly discovered in collections of English specimens. But the use of the term in museums and in books is such as to include a much greater variety of forms than seemed to Blainville and Voltz fairly admissible. Taking the idea of this Belemnite from the text and figures of Voltz, we have a cylindroidal, much compressed form, quite obtuse, and even-rounded to an egg-shape at the end; and upon making longitudinal sections the internal layers, after a certain age, correspond in bluntness of termination. The youngest layers end sharply. The cross section in the forward part of the alveolar region shows a remarkably thickened accretion of the sheath on the ventral aspect. There is always a ventral furrow in one variety, called *B. irregularis* by Schlotheim; and indeed the apex being sometimes umbilicate, sometimes papillary, and sometimes in other ways of unusual appearance, the term seems natural enough, though founded in part on accidents of decomposition rather than structure.

If now we turn to Quenstedt, and examine the figures given in pl. xxvi of his 'Cephalopoden' (1849), we find the ordinary blunt compressed form, with its various inner and outer aspects clearly represented (pl. xxvi, figs. 1—7.) Another is given, somewhat longer and less obtuse (pl. xxvi, fig. 8); again, one somewhat bowed at the end (pl. xxvi, fig. 11), called *B. digitalis acutus;* and a closely allied form (pl. xxvi, fig. 65), which shows traces of lateral furrows, receives the name of *B. incurvatus.* Finally, a longitudinal section (pl. xxvi, fig. 14 *c*), and two drawings of the lateral aspect of somewhat larger and more pointed specimens, with ventral and dorso-lateral grooves, complete the series of varieties under the title of *B. digitalis tripartitus.* Of all these Belemnites, which are found only in the Lias, I have seen examples from English localities. Only a very small number correspond strictly to *B. irregularis* of Schlotheim, but there is a great number and variety comprised within the limits of the additional forms of Quenstedt. The middle term of all the varieties agrees with some specimens which are frequent at Whitby, called *B. vulgaris* by Young and Bird. I possess only one specimen from Whitby at all approaching to *B. digitalis* in the extreme bluntness and irregularity of the end, which is obscurely marked by a ventral groove, and is eroded. It was taken from the Upper Lias of Saltwick.

BELEMNITES IRREGULARIS, *Schlotheim.*   Pl. XV, figs. 37, 39.

Reference.   *Belemnites digitalis.* 'Faure Biguet,' 1810 (as quoted by Blainville in 1827, but disallowed by D'Orbigny, 1860).

B. irregularis, Schl. 'Taschenbuch,' p. 70, pl. iii, fig. 2, 1813, and 'Petref.,' p. 48, No. 5, 1820.

B. digitalis, Blainv., p. 88, pl. iii, figs. 5, 6, 1827.

„   Voltz, p. 46, pl. ii, fig. 5, 1830.

B. irregularis, D'Orbigny, 'Pal. Fr. Cephal.,' p. 74, pl. iv, figs. 2—8, 1842.

B. digitalis et *B. digitalis irregularis*, Quenstedt, 'Ceph.,' p. 416, pl. xxvi, figs. 1—8, 1849.

GUARD.   Short, straight, cylindroidal, much compressed; apex oblique and very blunt in the adult, often marked by a tubercular and pitted surface, from which vermicular ridges and hollows radiate to a short distance. Apex in younger specimens often sub-mucronate, in very young examples acute. On most specimens there is one short ventral furrow near the apex; no dorso-lateral grooves at the apex.

Transverse sections elliptical, with flattened sides; across the anterior part of the alveolar region, the unequal thickness of the ventral and dorsal parts of the guard is often conspicuous; and, what is remarkable, in specimens which have the one short apicial furrow, it is on this side that the thickening is greatest. Such a groove usually indicates the ventral aspect, and it is so figured by Quenstedt ('Cephal.,' pl. xxvi, fig. 1 *d*). Voltz, who observed this unequal thickening, and paid much attention to it, states that, while from the apex toward the alveolar cavity the deposited layers thicken most of all the dorsal parts, and least of all the ventral, it is quite otherwise in the alveolar region, where the augmentation takes place on the ventral part (Voltz, 'Obs. sur les Bel.,' p. 47, and pl. ii, fig. 5 F″). He also remarks that the "summit" is oblique, and nearer to the ventral than to the dorsal aspect. I find, indeed, that it is on this side, toward which the apex is inclined, that the short apicial groove occurs, when it does occur, contrary to what is usual; and it is on this side that the circum-alveolar space is thickened. Before, however, deciding that Voltz is altogether right in this observation, I wish to obtain specimens from the Upper Lias of Gloucestershire, which may be properly cut for examination.

In a longitudinal section, the axis is seen to be very oblique, a little arched toward and near to the ventral side, and marked by a pale tint in the laminæ about it—probably they were less calcareous.

Proportions in the adult. Taking the dorso-ventral diameter at the alveolar apex at 100, the ventral part is 40, the dorsal part 60, the cross diameter 80, the axis 180 to 220.

PHRAGMOCONE. Oval in section, slightly arched toward the ventral region, and terminated by a spherule; the angle of inclination of the sides 25° according to Voltz, 20° to 22° according to D'Orbigny. Voltz describes the summit of the dorsal "ogives" as nearly rectangular. The diameters are 100 to 87.

VARIETIES. *a.* Short, with obtuse ovoid summit inclined toward the grooved side of the apex (Pl. XV, fig. 37).
β. Longer, with obtuse summit (Pl. XV, fig. 39).

Locality, var. *a,* short, obtuse. Frocester Hill (Sands), in the Cabinet of Bristol Institution, two specimens marked B 1, No. 28. Also two specimens from the same locality, in the Cabinet of Mr. Moore, one of them longer in proportion.

BELEMNITES REGULARIS, n. s. Pl. XV, fig. 38.

GUARD. Straight, cylindroidal, tapering in a curve to the end, compressed, with an oval or elliptical section, and a three-grooved, somewhat striated, umbilicated termination;

ventral aspect rather narrower than the dorsal (adult) ; no recurvation of the apex, which is divided by the grooves.

Sections across the alveolar cavity show the slightly oval alveolar section within the oval or elliptical outline of the guard—the oval outline being caused by the contraction of the ventral region ; sides flattened.

Dimensions. Length of guard before it shows any alveolar expansion, $2\frac{1}{3}$ inches, of which the axis is about $1\frac{1}{2}$ inch ; the diameter from back to front being in the largest specimen $\frac{7}{10}$ths inch.

Proportions. The dorso-ventral diameter at the apex of alveolus being taken at 100, the cross diameter is 85, the ventral radius about 40, the dorsal about 60, the axis 250 (in the larger variety 300 to 400).

Locality. Upper Lias Clays at Eydon (one), and Badby (three), near Banbury (*Stuttard*).

Observations. The affinity of this tripartite fossil to that so frequent at Whitby (*B. vulgaris*) is obvious ; while its parallel sides and greater compression offer analogies to the less obtuse examples of *B. digitalis*, like Quenstedt, pl. xxvi, fig. 11, and *B. incurvatus*, fig. 15, on the same plate.

As in *B. vulgaris*, so in this, the axis of the guard varies in length in proportion to the diameter. There is a series of more slender forms in Mr. Stuttard's collection from Eydon and Badby, in which the axis measures 320, 350, 400, no other difference being evident ; *e. g.* (Nos. 50, 51, 58), Fig. 38 *b* shows one of the shortest of these slender forms, with ventral groove longer and more distinct than usual.

Young specimens (Nos. 53 and 63) from Eydon and Badby have the axis remarkably excentric, in the proportion 28 to 72, the interior nearly circular, sides somewhat channelled, termination acute, ventral sulcus distinct, the others variable ; in one, dorsal plaits. The axis exceeds 600. They resemble almost exactly some young cylindroid varieties of *B. subaduncatus*, with the same excentricity. The older specimens referred to also correspond very much with the older cylindroid specimens of *B. subaduncatus*. Thus they constitute a parallel series ; and taking the whole into account, we have this general comparison of these tripartite forms :

Whitby, more or less recurved :

          Short form, *B. vulgaris*, Pl. XVI, fig. 40.

          Long form, *B. vulgaris*, Pl. XVI, fig. 41.

            ,,   *B. subaduncatus*, Pl. XI, fig. 29.

Banbury, little or not at all recurved :

          Short form, *B. regularis*, Pl. XV, fig. 38 *a*.

          Long form   ,,   ,,   fig. 38 *b*.

BELEMNITES VULGARIS, *Young* and *Bird*. Pl. XVI, figs. 40, 41.

Reference. Belemnites *vulgaris*, Young and Bird, Yorkshire Coast, p. 258, pl. xiv,
fig. 1, 1st ed., 1822 ; 2nd ed., p. 275, pl. xv, fig. 1,
1828.

,, ,, Simpson, 'Lias Fossils,' No. 23, p. 28, 1855.

B. curtus, Simpson, 'Lias Fossils,' No. 24, p. 29, 1849.

B. incurvatus, Quenstedt, 'Cephal.,' p. 418, pl. xxvi, fig. 15.

GUARD. Short, compressed, tapering in a curve to the apex, which in perfect young specimens is acute, but in older examples always somewhat obtuse; lateral furrows distinct for short distance only, ventral furrow present and distinct for a short space in one variety, obscure or almost obsolete in another.

Transverse sections show the contour to be more variable in different parts of the guard than is usual. At the apex of the alveolus the outline is oval, with more or less flattened sides somewhat inclined to one another, the ventral breadth being less than the dorsal; the axis excentric. Toward the apex the ventral region increases in amplitude, and the contour changes still more toward the apex, there showing three emarginations corresponding to the three grooves. This gives on the whole an irregularity to the surface-curves which is less manifest in other species, but is nevertheless often to be recognised in some of them.

Longitudinal sections show the form to have been little, if at all, changed by growth; the inner outlines, down to ½ an inch of axis of guard, giving nearly the same proportions as the outer surface of a specimen with four times as long an axis.

Greatest length observed 6 inches, including the expanded part of the guard; greatest length of axis, commonly under 2 inches.

Proportions. Taking the dorso-ventral diameter at the alveolar apex at 100, the ventral radius is about 45, the dorsal about 55, the cross diameter 80, the axis usually 200, 225, 250, 275.

Young. I have found it difficult to trace this species through its younger forms by selection of specimens on the Scars of Whitby and Saltwick. It appears, however, by sections to vary but little with age. In my youngest specimen the axis of the guard is less than 1 inch long, the normal diameter ⅜ths—proportions which are also found in older specimens.

PHRAGMOCONE. Oblique, a little arched, with an elliptical section; sides converging 19° in the anterior parts, and 22° near the apex. The axis of the phragmocone in two specimens is more than twice as long as the axis of the guard. The diameters are as 95 to 100, or in some examples almost equal. The depth of the inter-septal spaces is commonly ⅓th of the long diameter near the apex; but in one specimen at a more advanced part of the cone it is $\frac{1}{10}$th.   The conotheca is finely striated and broadly undulated across the ventral region, longitudinally striated on the dorsal region, and there marked by a slightly prominent mesial linear band.   Fifty-five septa were counted in the space of half an inch from the apex of the phragmocone.

Observations.   This very common Belemnite varies as to the apex, which in young examples is acute, or even a little produced, while in old specimens it is always obtuse; there is always a tendency to greater curvature toward the apex in the ventral than in the dorsal region; the distinctness of the grooves varies; especially this is the case with the ventral groove, which is occasionally almost obsolete.   Simpson marks two varieties, viz.:

> *a.* Two grooves and numerous fine striæ at the apex.
> *b.* Three distinct grooves at the apex.

These varieties occur together in the Upper Lias near Whitby; I find the striation to occur in both in good young specimens; the differences as to the ventral groove are such as to unite the whole into one group, though in that particular the extreme examples appear distinct enough.   The apex is often eroded, so as to be, in the sense of Blainville, umbilicate.

The account of this Belemnite given by Young and Bird requires some elucidation. In the first edition of their work it is illustrated by two figures, one (pl. v, fig. 6) representing the guard engaged in a mass of Pentacrinites from the lower **part** of the Upper Lias—a fair specimen of Mr. Bird's talents as an artist; the other (pl. xiv, fig. 1) not so well drawn, but showing the exserted phragmocone.   These two figures represent two distinct forms—distinct varieties at least.   In the second edition, only one figure is given (pl. xv, fig. 1), and that not so good as either of those previously published.   In the text of both editions a careful description is given; one of the references is to *B. elongatus* of Miller, certainly an error.   The length is said to reach 10 or 12 inches, with a diameter of 2 inches at the broad end, but this must be very unusual.

Locality.   Whitby, Saltwick, Sandsend, and other places in the Upper Lias of Yorkshire, at various stages above the Jet-beds, up to and including the Leda-beds (*Young* and *Bird, Simpson, Phillips*).   These show two varieties as to length, the axis of the guard being proportionately shorter in fig. 40, and longer in fig. 41.

BELEMNITES RUDIS, n. s.   Pl. XVI, fig. 42.

GUARD.   Short, somewhat compressed, and oval in section, with or without one short obscure apical groove; no dorso-lateral grooves.   Adult specimens blunt at the end.

Sections across the alveolar region (Pl. XVI, fig. *s'*) show the slightly oval outline of the phragmocone within the slightly oval guard, which is thickened on the dorsal aspect; those across the post-alveolar region show the axis to be very excentric till near the apex (Pl. XVI, figs. *s″ s‴*).

Longitudinal sections give a nearly straight axis, prolonged in the direction of the ventral line of the phragmocone.

Proportions. Taking the dorso-ventral diameter at the alveolar apex at 100, the ventral part is less than 40, the dorsal more than 60, the cross diameter 95, and the axis 180 to 200 (in young specimens 300). Between the apex of the phragmocone and the apex of the guard the dorsal radius is sometimes twice as long as the ventral.

PHRAGMOCONE. Somewhat oval in section, slightly arched toward the ventral side; angle 28°.

Locality. Very abundant in ironstone-beds belonging to Middle Lias, between the jet-rock and marlstone, east of Staithes, Yorkshire (*Phillips*).

ON A GROUP OF BELEMNITES ALLIED TO *BELEMNITES COMPRESSUS* OF BLAINVILLE AND VOLTZ.

The title "*Belemnites compressus*" appears to have been first used by Stahl, as already noticed, p. 41. But the papers of this author in the 'Correspondenzblatt der Würt. Landw. Vereins,' 1824, attracted little notice, and the greater number of Belemnitologists followed the example of Blainville, 1827, who gave the name of "*compressus*" to a different species, belonging to a different group. In his description the essential points are as follows :

Shell straight, very compressed, so that the vertical diameter is much greater than the transverse, and the section is oval; apex medial, straight, with a broad shallow groove on each side, dying out by degrees toward the alveolar region; alveolar cavity conical, with an oval section, six inches and more long. Pl. ii, figs. 9 and 9 *a*, of Blainville's work represent the Belemnite laterally and in cross section. From these figures I infer that the axis of the guard was above three times as long as the diameter at the apex of the phragmocone. Such a specimen may be found in the sandy beds at the base of the Lower Oolite of Yorkshire, while in a higher calcareous rock occurs another and allied form, with five furrows at the apex, called by Blainville *B. quinquesulcatus*, and represented in his work, pl. ii, figs. 8, 8 *a*, 8 *b*.

Voltz, writing in 1830, gives a general character of *B. compressus*, and places under this title three varieties. The general characters are these :

Sheath large, straight, conoidal or conical, compressed, with an oval cross section in the alveolar region. Summit straight, *emoussé*, furnished with two dorso-lateral furrows, which pass over at least one third of the apical region. Axis of the guard excentric,

11

approaching the ventral side at its origin, and then bending toward the back and becoming subcentral. Phragmocone depressed, inclined toward the ventral side. The three varieties are—

Var. A.   *B. compressus*, Blainville, pl. ii, fig. 9 (Sand below Inferior Oolite).
  „   B.   *B. compressus*, Voltz, pl. v, fig. 2 (Upper Lias).
  „   C.   *B. compressus*, Voltz, pl. v, fig. 1 (Upper Lias).

The varieties B and C are fully described. The former has a perfectly conical sheath, much striated at the point, the striæ continuing on the shell nearly as long as the grooves, viz. for half the length of the apicial region. In the latter variety the sheath is conoidal, and the striæ are described as plaits about ten in number. The phragmocone in var. B is straight, the sides meeting at an angle of 26°; in C it is a little arched, and the sides meet at an angle of 29°. Ventral furrows were not observed in any of the specimens, which, except in this particular, appear very similar to examples from the Upper Lias of Yorkshire.

Quenstedt (1848) has treated this perplexing subject with attention. He employs the title of *B. compressus* for the fossils described by Voltz, from which he separates those of Blainville, and uses such compound terms as *B. compressus gigas*, *B. compressus paxillosus*, and *B. compressus conicus*, for allied forms of the same natural group.

We have in the English Lower Oolites and Upper Lias plenty of examples of this group of "compressed Belemnites," but they have not been strictly studied and compared.

Mr. James de Carle Sowerby, in 1829, represented the fossils from the Oolite of White Nab, near Scarborough ('Min. Conch.,' pl. 590, fig. 4), under the name of *B. compressus* of Blainville. Professor Morris, in his 'Catalogue of British Fossils' (1854), refers the same specimens to *B. giganteus* of Schlotheim. Since that date D'Orbigny (1860) has collected under the same title *B. ellipticus* of Miller, *B. compressus* of Sowerby, *B. quinquesulcatus* and *gigas* of Blainville, *B. gladius* of Deshayes, *B. Aalensis* and *longus* of Voltz, *B. grandis* and *acuminatus* of Schubler, *B. bipartitus* and *canaliculatus*, of Hartmann; but this is not a method to be recommended. This is not a species, but a group of species, whose geological range includes the Upper Lias and the Lower Oolite.

Under *B. compressus* of Blainville D'Orbigny ranged also *B. apicicurvatus* and *B. bicanaliculatus* of that author, *B. crassus* of Voltz, and *B. bisulcatus* and *B. tumidus* of Zieten.

It appears to me that three distinct British forms may be well marked among the varieties of Belemnites properly referred to *B. compressus* of Blainville and Voltz; one is described by the first author; another includes the varieties B and C of Voltz; the third is now illustrated from Yorkshire specimens.

BELEMNITES VOLTZII, n. s.   Pl. XVII, fig. 43.

Reference.   *Belemnites compressus*, Voltz, var. B and C, ' Belemn.,' p. 53, pl. v, figs. 1 and 2, 1830.

GUARD.   Conoidal, compressed, smooth, with a central blunt or eroded termination (when old).   Two dorso-lateral furrows proceed along half the space of the apicial region, and from the termination about ten small plaits extend to half the length of the furrows. No ventral sulcus.

Transverse sections show at the alveolar apex a regularly oval contour, the innermost layers of the guard undulated by the dorso-lateral furrows.

Longitudinal sections show the axis to be excentric, most so at the alveolar apex, from which it is reflected toward the back, and then continues subcentral to the end. The inner and younger laminæ end bluntly in arches, so as to indicate the young forms to have been obtuse at the termination (Voltz made this remark, page 55 of his work).

Greatest length observed, 5 inches, the diameter at the extremity being 1·2 inch.

Proportions.   Taking the diameter from back to front at the alveolar apex at 100, the cross diameter is about 90, the ventral radius 40, the dorsal 60.   The axis in variety B, Voltz (shorter and smaller variety), is 250 ; in the longer and larger variety, C, 350.

PHRAGMOCONE.   Nearly straight in var. B, slightly arched in var. C ; in var. B the dorsal region occupies one sixth of the circumference, in var. C one fourth ; in B the hyperbolar regions occupy the twelfth of the circumference, in C one eighth.   In var. C, along the band which separates the hyperbolar from the ventral region occur small striæ, which cross the hyperbolic arcs.   Of these Voltz says he could give no explanation.   (See *B. inornatus.*)

Locality.   I am not sure that specimens exactly corresponding with this description, and constantly deficient of the ventral sulcus, have come under my notice from any locality in England.   One cause of this doubt is the uncertainty about the younger forms; for while Voltz infers, from longitudinal sections, that the apex was always blunt in var. B, and that in var. C it became blunt with age, we find Quenstedt referring to figures of the young which are quite acute, with a very deep alveolus and very short axis (' Cephalopoden,' p. 422, pl. xxvii, figs. 13, 17).   But as he speaks also of the early appearance of the dorsal furrows and of the later appearance of the ventral furrows, "which sometimes become extraordinarily deep," it would seem to be the form here called *ventralis* which he is describing.

On the Scars at Whitby and Saltwick are many Belemnites which might be thought the young of the Voltzian shells, being quite devoid of ventral sulcus, but they are commonly acute, and not striated; others, which are acute and striated, with a ventral sulcus, may be thought to be the young of *B. ventralis*.

BELEMNITES VENTRALIS, n. s.   Pl. XVII, figs. 44, 45.

GUARD.   Conoidal, compressed, smooth, with a central blunt or eroded termination (when old).   Two dorso-lateral furrows and one medio-ventral groove proceed along nearly half the space between the end of the guard and the end of the alveolus.   The termination is striated, and the striæ extend to about half the length of the grooves.

Transverse sections at the apex of the alveolus show the axis excentric and the contour regularly oval; the ventral side rather narrower than the dorsal in the medial region, but wider in the part near the apex.   Further backward the section shows the effect of the ventral and lateral grooves.   The axis is nearly straight.   The greatest observed length, including about one third of the phragmocone, is 6 inches; the greatest diameter $1\frac{1}{4}$ inch.

Proportions.   The diameter ($v\,d$) at the apex of the alveolus being taken at 100, the ventral radius is 41, the dorsal 59, the axis 300 to 350, the cross diameter 90.

PHRAGMOCONE.   Slightly arched, ending in a spherule, with a slightly oval section (100 to 95 or 90); septa obliquely descending to the siphuncle, and waved on the edge; the depth of the chambers equal to one seventh of the diameter.   The sides are inclined to each other 21°.   The conothecal striation is very distinct.

Locality.   Upper Lias of Whitby, Saltwick, and Robin Hood's Bay (*Phillips*). Frocester Hill Sands (*Moore*).   Not observed at Lyme Regis (*Phillips*).

BELEMNITES INORNATUS, n. s.   Pl. XVIII, fig. 46.

Reference.   Belemnites compressus, Blainville, 'Belemn.,' p. 84, pl. ii, fig. 9, 1825.

GUARD.   Conoidal, compressed, smooth, with a blunt or eroded termination (when old).   Two dorso-lateral furrows proceed along one third of the apicial space.   No ventral furrow, no striæ about the termination.

Transverse sections at the apex of the alveolus give an excentric axis and a regularly oval contour, the ventral portion being rather narrower than the dorsal.   Toward the apex the lateral grooves indent the outline.

Greatest length observed, 4·65 inches, of which the axis is 2·65.

Proportions. The long diameter at the apex of the alveolus being taken at 100, the ventral radius is 42, the dorsal 58, the cross diameter 75 to 84, the axis 220, 250, 280. In an extreme case, a more lengthened specimen than usual, the axis is 360, and there is a further peculiarity to be noticed—a short dorsal groove.

PHRAGMOCONE. Somewhat remarkable for smoothness on the dorsal and dorso-lateral regions; for the ellipticity of the cross section, which augments toward the aperture, the axis appearing to be there in the proportion of 100 to 90; and for the small depth of the interseptal spaces, which are about one ninth of the diameter. In the first quarter of an inch from the apex are about 30 septa, in the second quarter 10, in the third quarter 7, in the fourth 5, in the fifth 4. The right and left sides of the phragmocone are straight, the ventral face very little concave, the dorsal very little convex.

The ventral portion of the conotheca, half the circumference, is undulated across by the convexity of the interseptal spaces, and finely striated both parallel to the undulations and lengthways. At the boundary between the ventral and lateral spaces these striæ turn up suddenly across the band (which is marked by two or three longitudinal straight lines), to join into and form the hyperbolic arcs, which are traced accurately to the asymptote. Along the band here referred to the striation appears complicated, and this probably is what Voltz refers to in his account of *B. compressus*, var. c, as already noticed. Beyond this the forward-bent arcs undulate the dorsal region, and are crossed by numerous longitudinal striæ, which are most distinct on the middle of the back.

VARIETIES. In general figure some specimens are nearly straight-sided, others are more convex in the middle of the apicial region, and more rounded toward the apex, which appears to be always blunt and central. The more convex the medio-apicial space, the shorter is the axis in comparison with the diameter. One of my specimens is very compressed, so that the cross section at the apex of the alveolus gives an ellipse of 100 to 75. The axis in this case measures 240.

One of my specimens from Blue Wick, which has the extreme length of axis already referred to, is remarkable for a short *dorsal* groove, corresponding in situation to that already noticed in *B. dorsalis*, p. 58, viz. the lower alveolar region.

Locality. In the Sandy Cap-beds of the Lias at Blue Wick, with *Vermicularia compressa* (*Bean, Phillips, Cullen*). In the Sands at Nailsworth (*Moore*).

Remarks. Though I have described these interesting fossils under three specific names, and think it most convenient to do so, it may possibly happen that more complete inquiry may lead me to regard them as varieties of one species, known as *B. compressus*, a name very appropriate, but already pre-engaged by Stahl. If this view should finally prevail, a new comprehensive name will be required, and I hope the title of "Voltzii" may in that case be accepted, as a just tribute to one of the best of Belemnitologists. We should then have—

Belemnites Voltzii, var. *conicus*, Pl. XVII, fig. 43.
        ,,   var. *ventralis*,   ,,   figs. 44, 45.
        ,,   var. *inornatus*, Pl. XVIII, fig. 46.

In coming to this conclusion we might be confirmed by some other instances of the variability of the ventral sulcus, and it appears to me very interesting and significant to notice the loss of this furrow in passing from the upper part of the Lias to the lower part of the Oolites, which is immediately superposed, while above these lower beds no Liassic forms recur, but a new series of Belemnites begins. In general form we may remark in some specimens convex outlines in the retral slopes of the ventral and dorsal faces near the apex, in others this same part is quite straight-sided, or even a little produced, so as to approach the figure of *B. longisulcatus* of Voltz (pl. vi, fig. 1). But a more important circumstance, already mentioned, is the presence or absence of the ventral groove. In the very uppermost part of the argillaceous Lias beds of Whitby, which contain *Leda ovum*, I found specimens all sulcated on the lower side; in the lowest sandy Oolitic Dogger of Blue Wick I found others not sulcated—this being the main difference which I observe. In a series of these Belemnites lately collected for me by Mr. Peter Cullen the same fact appears. On examination it appears that striation of the apex commonly accompanies the ventro-sulcate variety, and commonly is absent from the other. What appear to be rather young examples of both forms are frequent in all the Upper Lias beds of Whitby, but large specimens are rare there. The younger examples are very acute at the end, the older specimens commonly obtuse or worn or truncated, just as Voltz represents his examples from Gündershofen.

BELEMNITES LONGISULCATUS, *Voltz.*   Pl. XIX, fig. 47.

Reference.   *Belemnites longisulcatus*, Voltz, ' Belemn.,' p. 57, pl. vi, fig. 1, 1830.
                 B. acuarius longisulcatus, Quenstedt, 'Cephal.,' p. 412, pl. xxv, f. 23, 1849.

GUARD.   Conoidal, straight-sided, much elongated, much compressed, rounded at the end, with two dorso-lateral furrows, three dorsal, and five ventral plaits; the dorso-lateral furrows occupy more than half the length, the plaits one fourth of the length; axis subcentral.

Transverse section oval, axis but little excentric; taking the diameter at the alveolar apex as 100, the axis is 500 or 600.

PHRAGMOCONE. Straight-sided, angle 25°. Dorsal area less than one fourth, lateral area one eighth of the circumference. Septa approximate.

This description is mostly from Voltz, who had one complete guard to examine from the Upper Lias of Wurtemberg. That this species occurs in the Upper Lias, on the York-shire coast, I have no doubt, but no good specimens have come to my hands. I there-fore copy the figures of M. Voltz. It is decidedly analogous to *B. compressus* of that author, but longer. In the Museum at Whitby is a specimen, No. 454, which agrees in general with the description and figure of Voltz, but it has a very long apicial groove on the ventral aspect. It is named *B. inæquistriatus* by Mr. Simpson, and bears to *B. lon-gisulcatus* the same relation that *B. ventralis* bears to *B. compressus*, B and C, of Voltz. In the same Museum is a specimen called " *B. compressus*, Young and Bird," with many longitudinal striæ on the *flattened* apicial region. I made hasty sketches of these, and think one of them of sufficient interest to be here reproduced and described.

BELEMNITES INÆQUISTRIATUS, *Simpson*. Pl. XIX, fig. 48.

Reference. *Belemnites inæquistriatus*, Simpson, 'Lias Fossils,' No. 3, p. 24, 1855.
   B. acuarius tricanaliculatus, Quenstedt, 'Cephal.,' p. 414, pl. xxv, figs. 13, 14, 15, 1849.

GUARD. Conoidal, uniformly tapering, much compressed, much elongated, obtuse at the end, with two dorso-lateral furrows, one ventral furrow, and several striæ. The ventral furrow extends over more than half the apicial space, the dorso-lateral furrow over one fourth, and the striæ equal or exceed these in length. The furrows are narrow and very distinct, making the section tripartite near the apex. The axis is above six times as long as the diameter at the alveolar apex. Greatest length observed, 6 inches.

PHRAGMOCONE. Unknown.

Locality. Upper Lias of Whitby (*Simpson*).

Observations. Closely allied to this, if not identical, is *B. erosus* of Simpson ('Lias Fossils,' No. 5), of which a specimen is to be seen in the Whitby Museum (No. 56); and perhaps the same may be said both of *B. compressus* of that author ('Lias Fossils,' No. 2) and *B. concavus* ('Lias Fossils,' No. 4), but this I leave for further inquiry.

BELEMNITES SULCI-STYLUS, n. s. Pl. XIX, fig. 49.

Reference. *Belemnites acuarius macer*, Quenstedt, 'Cephal.,' p. 414, pl. xxv, fig. 21, 1849.

GUARD. Very compressed, suddenly contracted behind the alveolar region, thence extended in a slender cylindrical form backward, and marked with two long, conspicuous, lateral grooves.   (No striæ seen.)

In the transverse section of the sheath, immediately behind the alveolar region, the outline is oval, the dorsal aspect being widest. Ventral radius 45 , dorsal radius 55, transverse diameter 80, axis 850.

PHRAGMOCONE.   Unknown.

Locality.   Only one specimen has come to my knowledge,—from Nailsworth, in the sands which cap the Lias (*Mr. C. Moore's* Cabinet).

Remarks.   *Belemnites acuarius* is a name applied to many forms of Belemnite, one of which has been already referred to as allied to *B. tubularis*—by some thought identical with it.   Quenstedt's figure, now referred to, appears certainly to agree with the Nailsworth specimen, and with no other that I have seen from English localities.   Another figure of Quenstedt's may probably be regarded as belonging to the species (*B. acuarius longisulcatus*, 'Cephal.,' pl. xxv, fig. 11), and fig. 12 of the same plate, which is striated, may, perhaps, be added to this reference.   I shall be glad to hear of more specimens from the English Lias and Liassic Sands.

It appears necessary to mark this by a definite name—"*acuarius*" being already used in the sense of D'Orbigny, "*macer*" being also employed by Mayer.   I propose the epithet "*sulci-stylus*" for the smooth, compressed, grooved forms, with cylindroid extension of the guard, not capable of being *flattened* by pressure.   I regard this flattening as a mark of imperfect original calcification characteristic of a small number of these elongated Belemnites, chiefly of the Upper Lias.

BELEMNITES ELEGANS, *Simpson.*   Pl. XX, fig. 50.

Reference.   Simpson's ' Lias Belemnites,' No. 40, p. 31, 1855.

GUARD.   Compressed, subhastate ; cylindroidal in the post-alveolar region, tapering from thence to an extended, striated, pointed apex ; slight traces of short dorso-lateral grooves at the apex ; on some specimens long, lateral, shallow grooves.

Transverse section elliptical till near the apex, where it is obscurely trilobed ; axis a little excentric.

Longitudinal sections show the axis to be nearly straight, and the young to have had more uniformly tapering sides.

In young specimens the proportion of the axis appears to be nearly the same, or rather longer, but the general figure is more uniformly tapering.

Dimensions. The largest specimen from Robin Hood's Bay which I have seen is in the Whitby Museum, No. 967, 5·2 inches long, and 0 65 in diameter at the alveolar apex. I have some nearly as large from Robin Hood's Bay, and many others of lesser magnitude from Huntcliff. My smallest specimen from Huntcliff is 1·5 inch long; and between these extremes my collection contains many examples, some more compressed than others.

Proportions. The diameter at the alveolar apex from v to d being taken at 100, the cross diameter is about or above 90, the ventral radius 40 to 44, the dorsal 60 to 56, the axis 400. Near the apex the ventral and dorsal radii become equal.

PHRAGMOCONE. The alveolar cavity is empty in the only specimen I could afford for longitudinal section. The angle is about 28°, but near the apex 32°, the whole figure a little arched.

Locality. Toward the upper part of the Lower Lias, in Robin Hood's Bay, north side (*Simpson, Cullen*); in the same position under Huntcliff (*Phillips*). In the Marlstone beds of Staithes and Robin Hood's Bay is found a Belemnite much resembling this, though usually with more distinct dorso-lateral grooves, and less distinct special striæ. It must be ranked as of the same species, and does not, I believe, occur in any higher strata of the Yorkshire coasts.

Observations. Mr. Simpson, who first noticed this form of Belemnite, describes the section as circular; it is, however, rather elliptical. He has recognised the younger forms at Robin Hood's Bay (Nos. 971, 972), as I have also done at Huntcliff. One of the specimens bearing the same name in the Whitby Museum (No. 976) is now regarded by Mr. Simpson, who discovered it at Robin Hood's Bay, as a distinct species, with the name of *B. scabrosus.* It will be described immediately. Two others in that Museum (No. 458) show no apicial grooves. As already observed (page 57), this fossil is allied to *B. lævis,* —and, we may add, to *B. subtenuis* of Simpson and *B. tripartitus* of Schlotheim, though the apicial plaits and grooves are usually so faint or so short as to make this analogy less obvious. From *B. lævis* its attenuated apicial region may be regarded as distinctive. Specimens occur at Lyme Regis, referred by me to *B. nitidus* (p. 67), and figured as a short variety (Pl. XIII, fig. 34 *b*), which are much like this fossil, but have not the apicial striæ, and always present more or less of the double lateral grooving. Other specimens from Lyme, which are referred to *B. apicicurvatus*, also resemble *B. elegans*, except in the apicial part of the guard.

BELEMNITES SCABROSUS, n. s. (*Simpson, MS.*) Pl. XX, fig. 51.

GUARD. Elongate, slightly compressed, fusiform, tapering to a lengthened apex; one short ventral groove, two short lateral grooves, which extend into lateral facets;

alveolar region expanded, covered with rough granulations; post-alveolar region enlarged. The grooves are distinct for about half an inch each.

Only one specimen is known, viz. "No. 976," in the Whitby Museum, of which Pl. XX, fig. 51, is a sketch. The total length is $7\frac{1}{2}$ inches, the diameter at the apex of the phragmocone less than 0·4 inch, giving a proportion of axis (as 1800 to 100 diameter) such as occurs in *Belemnites clavatus* and other very long Belemnites. In the post-alveolar enlargement the diameter becomes 0·45 inch.

Locality. Obtained by Mr. Simpson from the upper part of the Lower Lias, on the north side of Robin Hood's Bay.

Observations. It is very possible, indeed very likely, that by further research it may be proved that this really elegant Belemnite is a full-grown example of one of the claviform shells which occur in the Lower Lias of Yorkshire. With a view to settle this and some other questions, I have examined four times the beds which yield these Belemnites at the base of Huntcliff, collecting many specimens; and Mr. P. Cullen has with equal care explored for me the corresponding beds of Robin Hood's Bay. At present, looking on a hundred specimens, I am not able to furnish evidence in favour of the opinion. Mr. Simpson formerly included it with less hastate forms in the species he called *B. elegans*, but the great proportionate length of its axis is probably decisive against that alliance. The rugulosity about the alveolar region may be of some importance as a character, and yet not really diagnostic, just as in *B. tubularis* another kind of rugulosity is frequently but not constantly found on the alveolar region.

BELEMNITES CYLINDRICUS, *Simpson*. Pl. XX, fig. 52, P. *l*, P. *v*, P. *d*.

Reference. *Belemnites cylindricus*, Simpson, 'Lias Fossils,' No. 27, 1855.

GUARD (old). Cylindrical in the alveolar region, tapering evenly to a rather blunt, often rather recurved summit, from which two distinct short dorso-lateral grooves proceed, and lose themselves before reaching the alveolar region. In a very perfect specimen the apex is striated on the ventral and on the dorsal surface; the ventral striæ are accompanied by a short groove, not seen in any other example.

Sections show a nearly circular contour, with the axis a little excentric and straight. The ventral face is sometimes a little flattened (ventro-planate, as in *B. ventroplanus* and *B. subdepressus* of Voltz).

Greatest length observed (including an expanded part), 6 inches, of which the axis of the guard occupies 4; diameter of the cylindrical part, $\frac{6}{10}$ths of an inch.

Young specimens are rarely seen, and are more pointed than old ones, and show hardly a trace of apicial grooves or striæ.

Proportions. The diameter (*v d*) being taken at 100, the ventral radius is 40, the dorsal 60, the cross diameter 100, the axis 420.

PHRAGMOCONE. Nearly straight, with a nearly circular section; the angle $m = 25°$.

Locality. In the lower part of the Upper Lias at Saltwick (*Phillips*); at Robin Hood's Bay (*Cullen*); in ironstone layers at Kettleness (*Simpson*); in the shale under the Jet-bed, plentifully; and in ironstone layers at Staithes and Rosedale (*Phillips*); in the Marlstone series below the ironstone.

Observations. The agreement of this Belemnite with that long known as *B. paxillosus* is obvious and intimate, and the resemblance of particular selected specimens is almost complete, the principal observable difference being a greater proportionate length of axis and a longer tapering to a less obtuse apex in the Yorkshire specimens.

For comparison, a specimen from Ilminster, in Mr. Moore's Cabinet, is represented fig. 52, *Ml*.

Recurvation of the apex occurs in several of the specimens of *B. paxillosus*, especially in those from Ilminster; in several of the specimens of *B. cylindricus* from Rosedale, near Staithes, it is so pronounced as to approach the form of *B. aduncatus*.

On the whole, I can hardly doubt that the Yorkshire specimens agree with *B. paxillosus amalthei* of Quenstedt ('Cephal.,' pl. xxiv, fig. 5); the state of conservation seldom allows of the striation of the apex to be perfectly seen, as in our representation of *B. paxillosus* (Pl. XX, fig. 52, *Ml*).

B. elongatus, *B. apicicurvatus*, *B. paxillosus*, and *B. cylindricus*, taken together, compose a natural group of generally cylindrical or cylindroid forms, with dorso-lateral grooves at the apex, and plaits or striæ on the ventral and dorsal aspects (exceptionally, a deeper stria on the ventral and also on the dorsal face). They are unknown in Lower Lias, but extend from the base of the Middle Lias to the lower part of the Upper Lias, and are found in Dorsetshire, Gloucestershire, Northamptonshire, Lincolnshire, and Yorkshire.

BELEMNITES OXYCONUS, *Quenstedt.* (Diagram, No. 23, p. 88.)

Reference. *Belemnites (tripartitus) oxyconus*, Quenstedt, 'Cephalop.,' p. 419, pl. xxvi, figs. 19, 20, 1849.

GUARD. Compressed, conoidal or conical, ending in a produced, pointed, somewhat reclined apex; lateral grooves extend over the alveolar region.

Transverse section oval, the ventral region broadest.

Locality.   Cheltenham, in the Belemnite-bed of the Lower Lias (*Buckman*).

DIAGRAM 23.

Belemnites oxyconus.

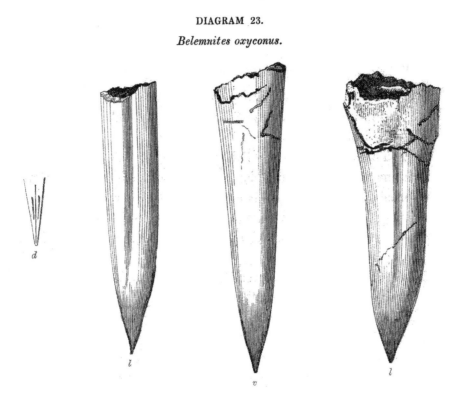

The fossil here represented, from Mr. Buckman's Collection, is from the "Belemnite-bed" of the Lower Lias of Cheltenham.   It corresponds to all appearance with the species referred to by Quenstedt, and I find it also to be much allied to *B. elegans* of Simpson (p. 84 ; and Pl. XX, fig. 50) ; but it is not striated about the point, as that species seems always to be ; its axis is shorter, and the figure is more oblique.   It seems allied to *B. acutus* of Miller, and may be the old form of that species.

EXPLANATION OF PLATE VIII.

FIG.

18. BELEMNITES BUCKLANDI, n. s.

    l'. Copy of the figure given by Dr. Buckland in his 'Bridgewater Treatise,' second edition, pl. lxi, fig. 7. In the upper part of the conotheca lies the ink-bag (opposite *i*); lower down, the phragmocone appears; and the figure is completed by the somewhat hastate guard. Lyme Regis. Collection of Miss Philpotts.

    l''. Specimen (seen laterally) from Blue Wick. Collection of Prof Phillips.

    l'''. Another, from the same locality, seen laterally.

    v'''. Ventral aspect of this specimen.

    v. Ventral aspect of a specimen which shows erosion along the face.

19. BELEMNITES MILLERI, n. s.

    l'. Lateral view of a specimen in the Oxford Museum.

    v'. Ventral aspect of the same.

    l''. Lateral view of a specimen in the Collection of Prof. Phillips.

    v''. Ventral aspect of the same.

    s'. Longitudinal section, showing the phragmocone. Oxford Museum.

    s''. Longitudinal section of a specimen in Prof. Phillips's Collection.

    s. Transverse section of the guard.

    a. Umbilicated apex of a specimen.

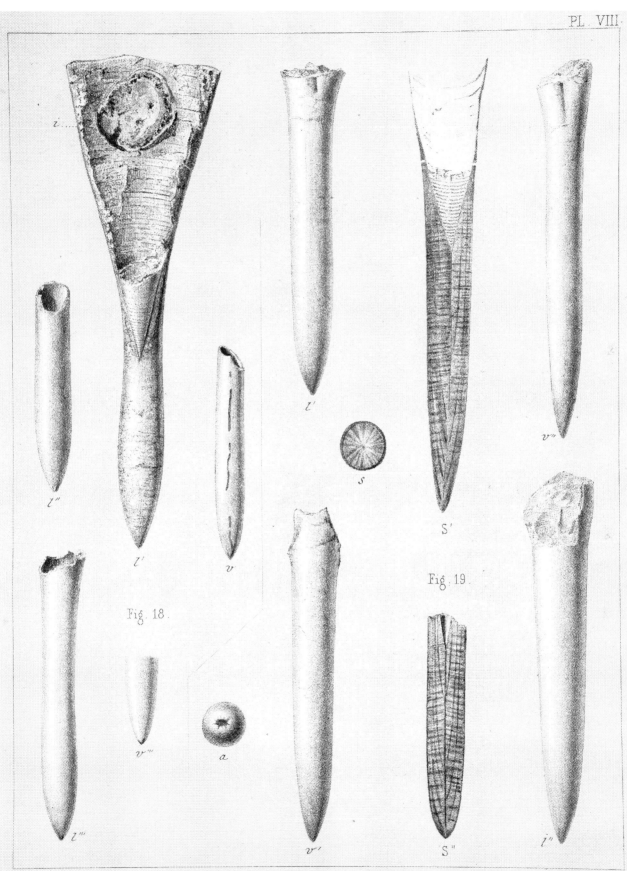

Imp. Becquet à Paris.

EXPLANATION OF PLATE IX.

Fig.

20. BELEMNITES POLLEX.

 l'. Sketch of a specimen in the Whitby Museum.   Lateral view.

 v'. Lower portion of the same, to show a broad irregular depression, and two stigmata.

 l''. Smaller specimen, in the Collection of Prof. Phillips.   Lateral view.

21. BELEMNITES ACUMINATUS (FERREUS).

 l. Side view of a specimen in the Whitby Museum.

 s. Transverse section, to show the nearly circular outline.

22. BELEMNITES ACUMINATUS.

 l. Side view of a specimen in the Whitby Museum.

 d. Dorsal aspect, near the apex.

 s. Cross section, to show the nearly circular outline.

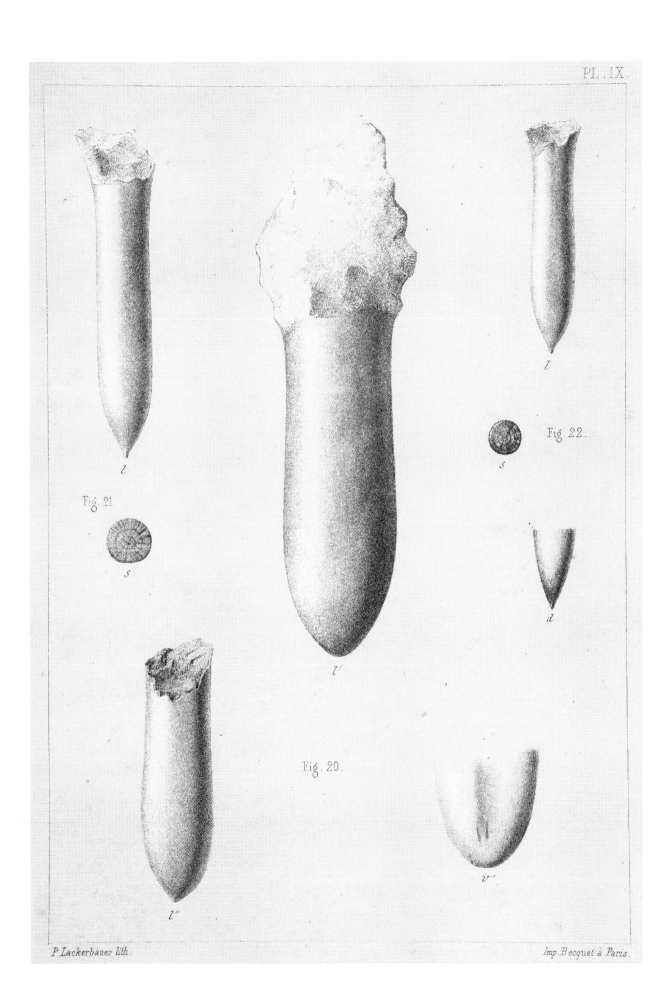

Fig. 22.

s

l

l

Fig. 21.

s

d

l'

Fig. 20.

l"

v'

P. Lackerbauer lith.

Imp. Becquet à Paris.

EXPLANATION OF PLATE X.

F<small>IG</small>.

23. B<small>ELEMNITES</small> L<small>ÆVIS</small>. (Whitby Museum.)

 l'. *Belemnites trivialis*—sketch from a specimen in the Whitby Museum. Lateral view.

 l''. *Belemnites lævis*—sketch from a specimen in the Whitby Museum. Lateral view.

 v. Ventral aspect, near the apex.

 d. Dorsal aspect, near the apex.

 σ'. Striæ, near the apex, magnified.

 σ''. Striæ broken into short plaits.

24. B<small>ELEMNITES</small> D<small>ORSALIS</small>, n. s. (Collection of Prof. Phillips.)

 l'. Lateral aspect, the specimen slightly angular.

 l''. Another specimen, seen laterally.

 d. A specimen seen dorsally, to show the short interrupted groove.

 σ. Striated apex ; not a common occurrence.

 s'. Transverse section near the alveolus.

 s''. Transverse section towards the apex.

25. B<small>ELEMNITES</small> S<small>TRIOLATUS</small>, n. s. (Whitby.)

 l. Specimen seen laterally, with fine striæ towards the entire apex.

 v. Specimen seen ventrally, with similar striæ.

 σ. The striæ, magnified.

 s. Transverse section.

26. B<small>ELEMNITES</small> L<small>ÆVIS</small>. (Whitby.)

 l. Specimen seen laterally, showing striation near the apex.

 v. The same, seen ventrally.

 s. Longitudinal section, showing arched alveolar cavity.

 φ. Section of phragmocone and spherule.

 s. Transverse section.

27. B<small>ELEMNITES</small> S<small>UBTENUIS</small>.

 l. Side view, showing one of the long faint dorso-lateral grooves.

 v. Ventral aspect, showing the faint ventral groove.

 d. Dorsal aspect, showing the two dorso-lateral grooves.

 s'. Section across the alveolus.

 s''. Section across the guard, in the middle.

 s'''. Section across the guard, near the apex.

 σ. Striæ, near the apex. The specimens were collected by Prof. Phillips at Saltwick, near Whitby, in Upper Lias.

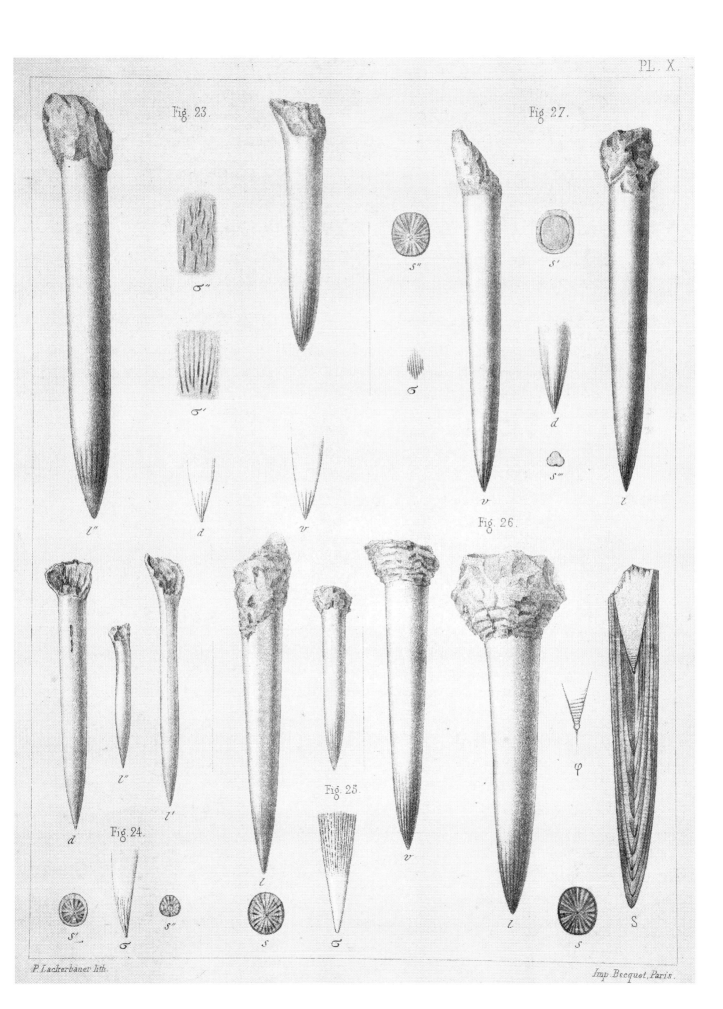

Fig. 23.

Fig 27.

Fig. 26.

Fig. 25.

Fig. 24.

P. Lackerbauer lith.

Imp. Becquet, Paris.

EXPLANATION OF PLATE XI.

Fig.

28. BELEMNITES TRIPARTITUS.   (Whitby.)

v.   Ventral aspect.

d.   Dorsal aspect, showing frequent striæ.

l.   Lateral aspect.   At v, d, is the apex of the alveolus.

s'.   Shows cross section at the alveolar apex.

s''.   The section near the apex.

σ.   The interrupted striation.   Specimen in Prof. Phillips's Collection.

29. BELEMNITES SUBADUNCATUS, in various stages of growth.   (Whitby.)

v'.   Ventral aspect of a young specimen.

l'.   Lateral aspect of the same.

d.   Dorsal aspect of a very young individual.

l''.   Lateral aspect of a more advanced and more hastate variety.

l'''.   The same aspect of one still older.

$d.'''$ Dorsal aspect of the same specimen.

l^{iv}.   The same aspect of one fully grown.

s', s'', s''', s^{iv}.   Transverse sections : s' being across the alveolar cavity; s'', at the alveolar apex; s''', towards the apex, unusually excentric; s^{iv}, near the apex.

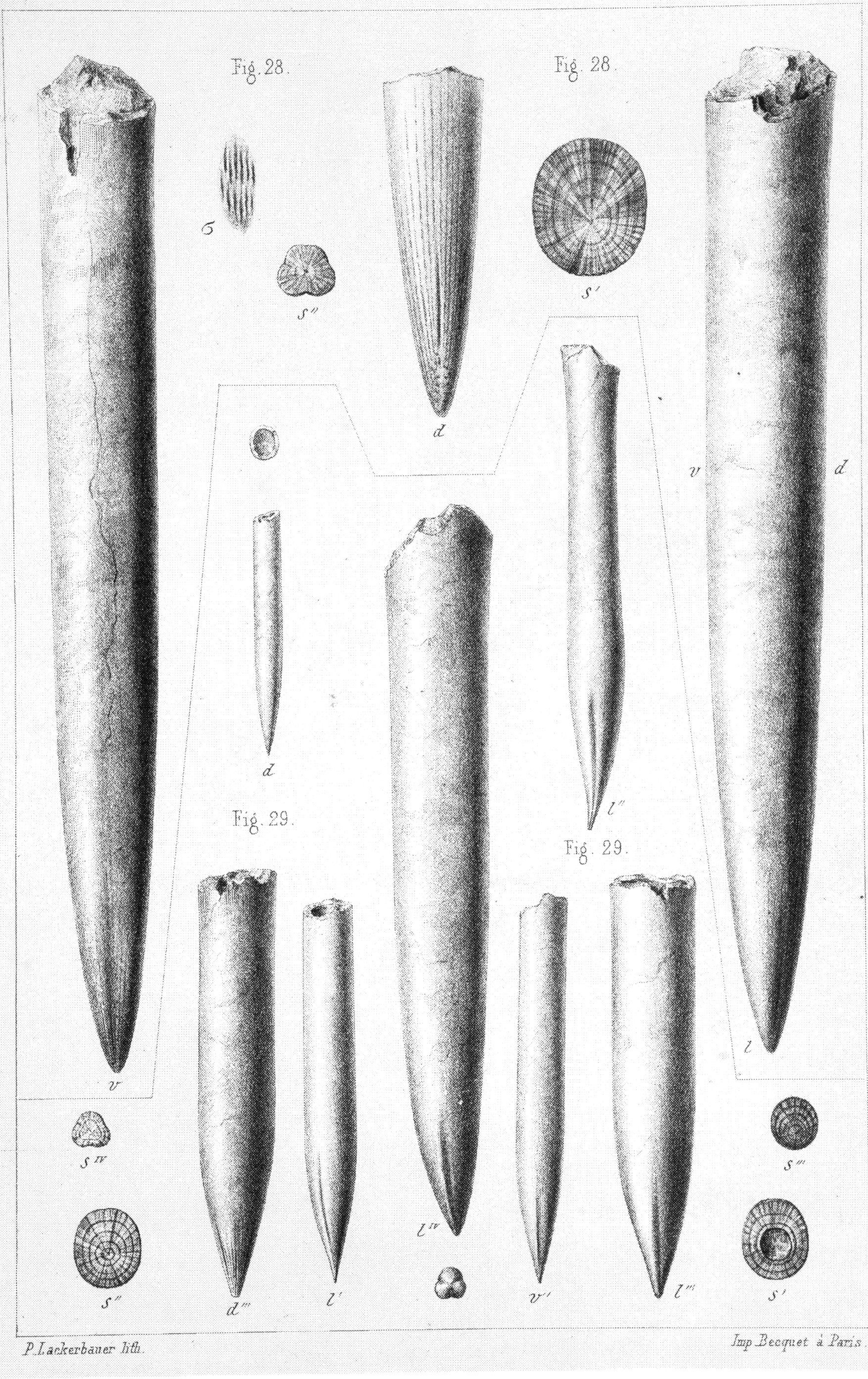

Fig. 28.

Fig. 28.

Fig. 29.

Fig. 29.

P.Lackerbauer lith.

Imp.Becquet à Paris.

EXPLANATION OF PLATE XII.

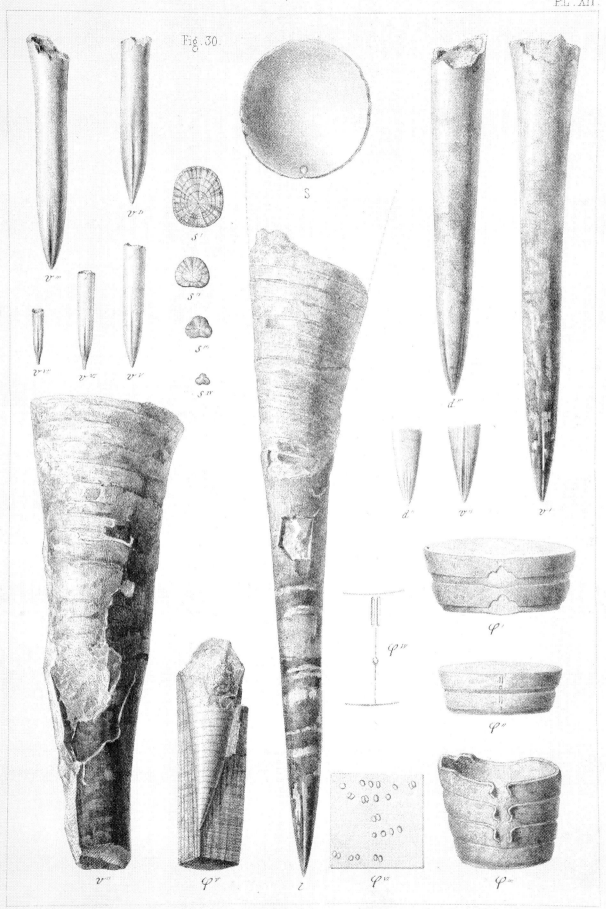

Fig. 30.

EXPLANATION OF PLATE XIII.

Fig.

31. BELEMNITES MICROSTYLUS, n. s.

    G′. Specimen belonging to the Geological Survey, and preserved in the Museum ("No. 349"), Jermyn Street. Lyme Regis.

    G″. Magnified view of the phragmocone and the enveloping sheath.

    O. Specimen in the Oxford Museum (part of Mr. Murley's Collection); from the Upper Lias of Dumbleton. The apex is broken off.

32. BELEMNITES LONGISSIMUS. (Lyme Regis.)

    l. Seen laterally.

    σ. The striation at the apex.

    $s.′$ Transverse section.

33. BELEMNITES JUNCEUS, n. s. (Lyme Regis.)

    The transverse sections ($s, s′$) show concentric sheaths, with undulations corresponding to the grooves; radiating fibres obscure.

34 a. BELEMNITES NITIDUS, n. s. (Lyme Regis.) Lateral views (l).

    The transverse section ($s′$) shows the lateral undulations and almost central axis.

34 b. A shorter variety, which has some affinity to *B. apicicurvatus.* Lyme Regis.

35. BELEMNITES QUADRICANALICULATUS. (Upper Lias [Sands], Chidcock.) Specimens in the Museum of the Geological Survey, Jermyn Street. Mr. Moore's Collection contains examples from the Upper Lias of Ilminster.

    $v, v′$, Ventral; $d, d′$, dorsal; $l, l′$, lateral.

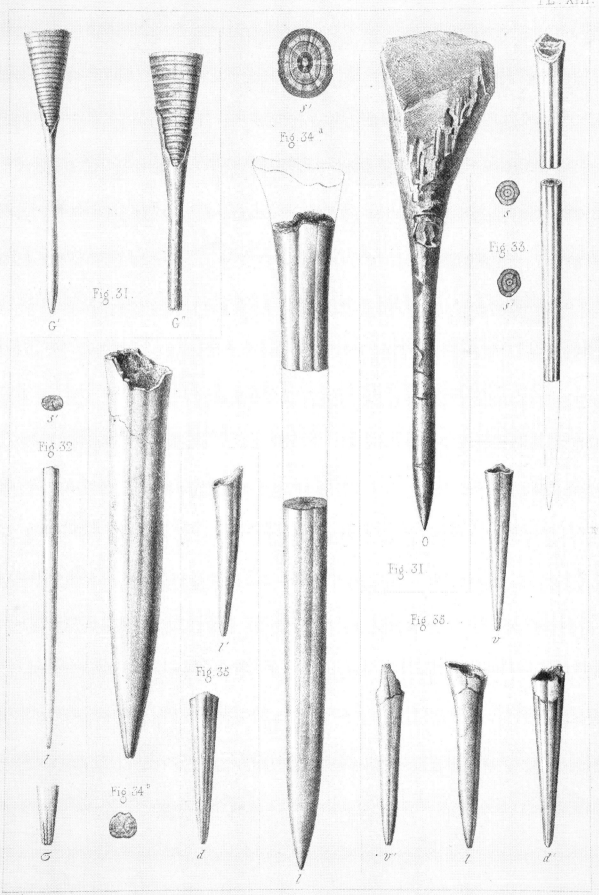

Fig. 31.

Fig. 34.ᵃ

Fig. 33.

Fig. 32.

Fig. 31

Fig. 35

Fig. 35.

Fig. 34.ᵇ

P. Lackerbauer lith.   Imp. Becquet à Paris.

Fig.

36. BELEMNITES TUBULARIS; the crushed part very long. The specimens are usually presented with a lateral aspect.

b'. Specimen from the Bristol Museum, marked, "B 1, 42. Near Gloucester."

b". Striation on the anterior part of the guard.

b"'. Crushed portion of the guard.

p'. Specimen from Whitby, belonging to Prof. Phillips. The crushed part comparatively short; the anterior part of the guard slightly furrowed; the aspect is lateral.

p". Opposite view of the same, showing a long dorso-lateral groove.

p"'. Ventral aspect of the same.

p . Transverse section at the alveolar apex.

p^v. Another specimen from Whitby; presented ventrally, showing some effect of pressure on the alveolar portion of the guard.

p^{vi}. Opposite aspect of the same.

p^{vii}. Lateral view of the same, showing the striation.

Fig. 36.

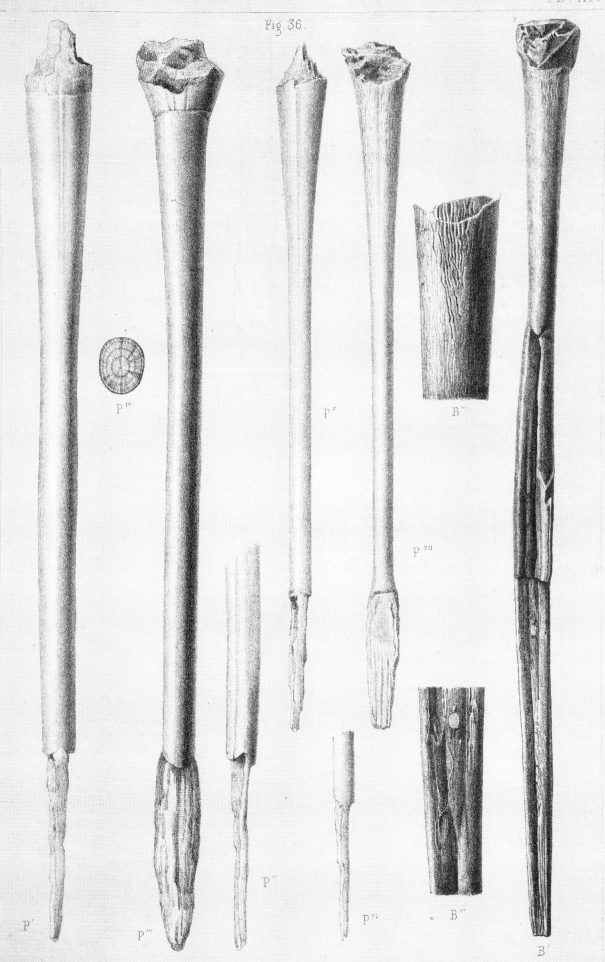

P' IV

P V

B ''

P VII

P'

P '''

P ''

P VI

B ''

B ''

B '

P. Lackerbauer lith.

Imp Becquet a Paris.

EXPLANATION OF PLATE XV.

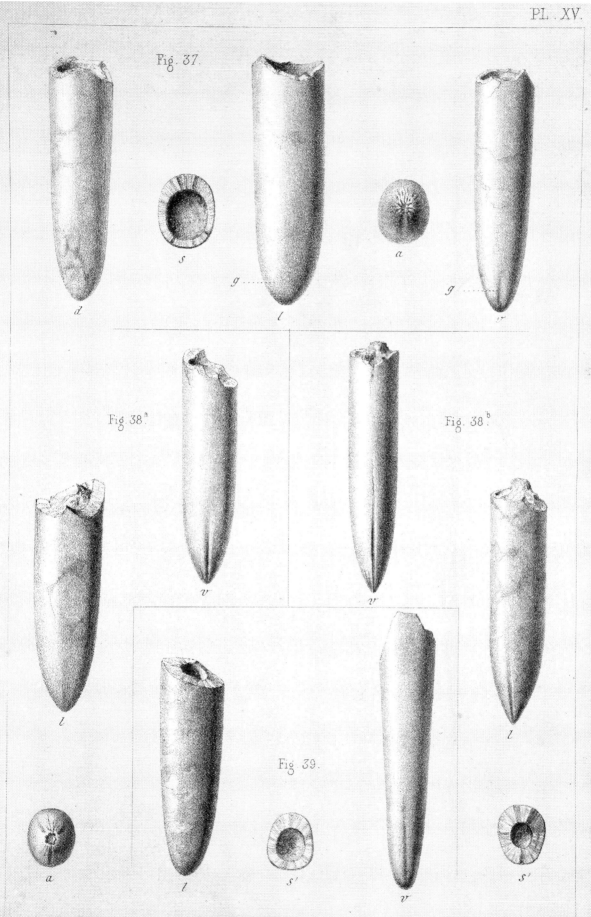

Fig. 37.

Fig. 38.ᵃ

Fig. 38.ᵇ

Fig. 39.

P. Lackerbauer lith.

Imp. Becquet à Paris.

EXPLANATION OF PLATE XVI.

<small>Fig.</small>

40. B<small>ELEMNITES</small> <small>VULGARIS</small>.   (From the Upper Lias of Whitby.)

  l.   Lateral and (*v*) ventral aspects of an ordinary middle-aged specimen.

  s′.   Section across the alveolar cavity ; in this case the ventral wall is thickest.

  s″.   Section at the alveolar apex.

  l′.   Lateral view of a young example.

  s.   Longitudinal section.

  φ.   Phragmocone, *in situ*, of a large specimen, with short axis of guard.

41. Lateral view of a somewhat longer individual.

42. B<small>ELEMNITES</small> <small>RUDIS</small>, n. s.   (From Staithes, near Whitby.)

  l.   Lateral view of an ordinary specimen ; the upper outline is supposed to be the terminal edge. *v* and *d* mark the ventral and dorsal aspect at the alveolar apex.

  v.   Ventral aspect of the same specimen, showing the obtuse and irregular groove.

  l′ and *d′.* Lateral and dorsal views of a young example, more lengthened than in the seniors.

  s′.   Section across the alveolar chamber.

  s″.   Section at the alveolar apex, showing great excentricity.

  s‴. Section nearer the apex of guard, showing less excentricity of axis.

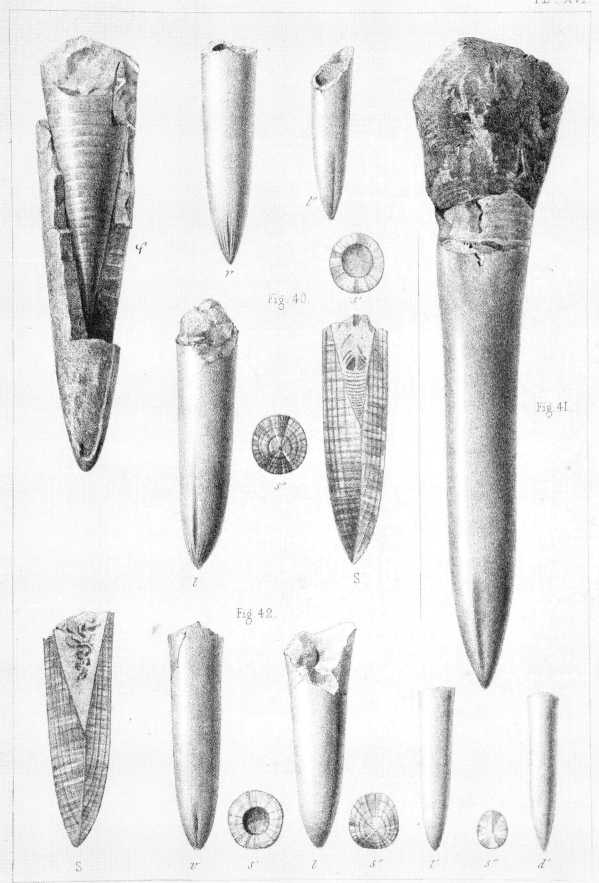

Fig. 40.

Fig. 41.

Fig. 42.

P. Lackerbauer lith

Imp. Becquet à Paris

EXPLANATION OF PLATE XVII.

43. BELEMNITES VOLTZII, n. s.   (Upper Lias, Whitby.)

   l.   Lateral view of a symmetrical specimen.
   d.   Dorsal aspect of the same.
   v.   Ventral aspect of the same.
   l″.   Lateral view of a larger specimen.
   φ′.   Phragmocone, seen laterally.
   φ″.   The same, seen dorsally.
   s′.   Transverse section at the alveolar apex.
   s″.   ,,   ,,   near the apex of the guard.

44. BELEMNITES VENTRALIS, n. s.   (Upper Lias, Robin Hood's Bay.)

   l.   Lateral view ; the apex worn.
   d.   Dorsal view of the same.
   v.   Ventral aspect of the same.
           The striation is seen on each.
   s′.   Transverse section at the alveolar apex.
   s″.   ,,   ,,   further along the guard.
   s‴.   ,,   ,,   near the apex of the guard.

45. Younger, examples, with unworn apices.   The points are striated.   (Upper Lias, Whitby.)

   l.   Lateral view.
   v′.   Ventral view, to show the rather long groove.
   v″.   Another specimen, in which only the faintest trace of such a groove appears.
           These three figures are not very easily distinguished from the young of
           Belemnites vulgaris.

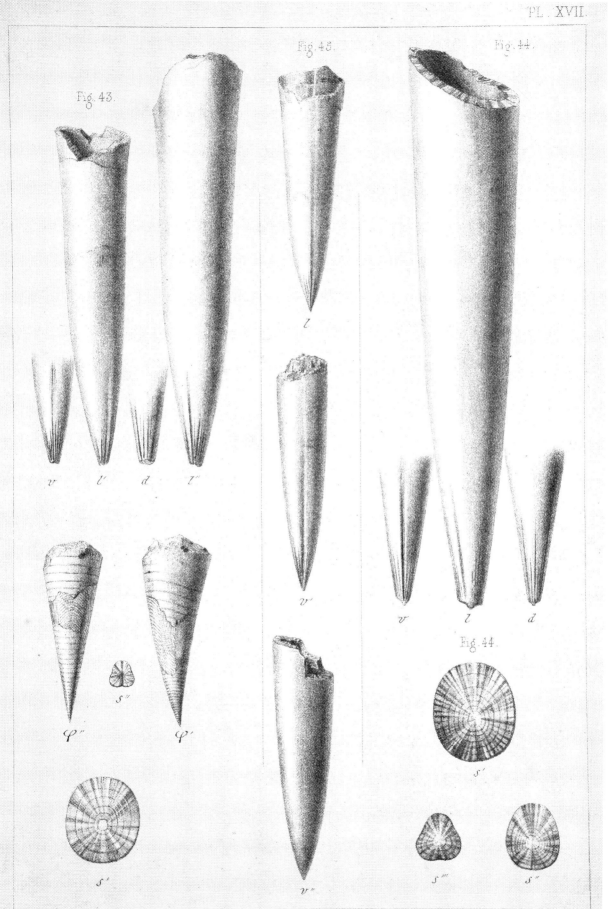

Fig. 43.

Fig. 45.

Fig. 44.

l

v   l'   d   l"

v'

v   l   d

Fig. 44.

s"

φ"   φ'

s'

v"

s'''   s"

P. Lackerbauer lith.

Imp. Becquet à Paris.

EXPLANATION OF PLATE XVIII.

Fɪɢ.

46. Belemnites inornatus, n. s. [Without ventral groove.] Blue Wick, Robin Hood's Bay.

l'. Lateral view, showing the insertion of the phragmocone. Between d and v is the alveolar apex.

l''. Lateral view of a specimen in which the grooves and striæ are very obscure (worn).

s. Longitudinal section, showing the slightly arched phragmocone, covered by the conotheca, *in situ*.

s'. Transverse section across the alveolar cavity.

s''.   ,,   ,, at the alveolar apex.

φ'. Portion of the conotheca, seen dorsally.

φ''. Portion of the same, seen laterally.

Fig. 46.

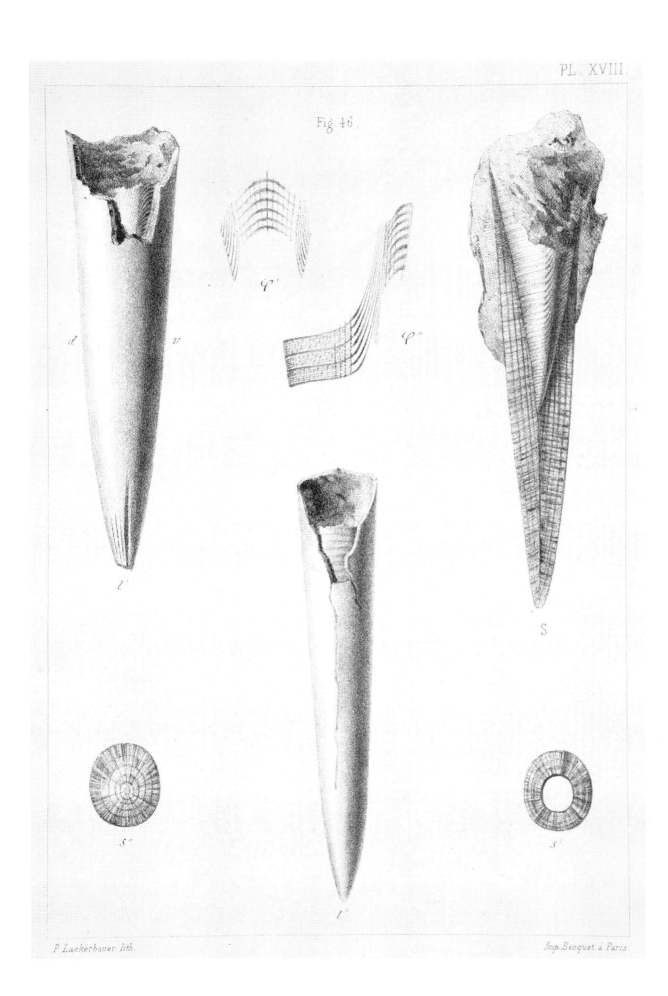

P. Lackerbauer lith.

Imp. Becquet à Paris.

EXPLANATION OF PLATE XIX.

Fig.

47. BELEMNITES LONGISULCATUS. (Copied from Voltz's figure, 'Obser. sur les Bélemn.,' pl. vi, fig. 1.)

    l. Lateral view.

    d. Dorsal view.

    v. Ventral view.

48. BELEMNITES INÆQUISTRIATUS. (From the Upper Lias of Whitby.)

    l. Lateral view of a specimen in the Whitby Museum, " No. 454."

    d. Dorsal aspect of the same.

    v. Ventral aspect of the same.

    v'. Enlarged view of the tripartite end.

    s'. Shows the oval contour of the alveolar region, the ventral area being broadest.

49. BELEMNITES SULCI-STYLUS, n. s. (From Nailsworth [Sands].)

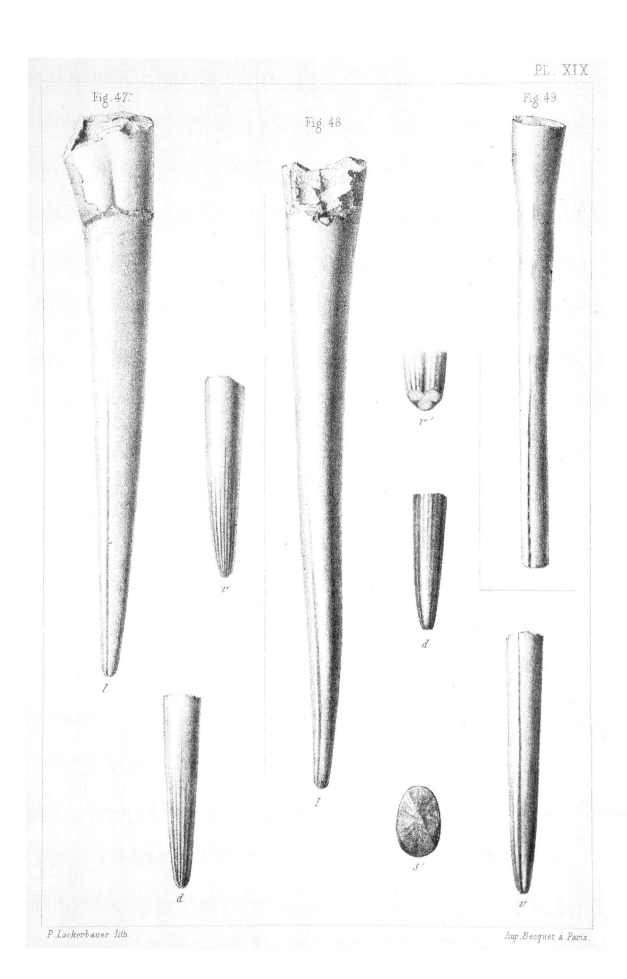

Fig. 47.

Fig 48

Fig 49.

v

l

d

v'

d

l

s'

v

P. Lackerbauer lith.

Imp. Becquet à Paris.

EXPLANATION OF PLATE XX.

Fɪɢ.

50. Belemnites elegans.

    l. Lateral view of a specimen in the Whitby Museum, "No. 967, Robin Hood's Bay, north side."

    l'. Lateral view of a specimen in Collection of Prof. Phillips. Huntcliff.

d, v. Dorsal and ventral aspect of the same specimen.

    l''. Lateral view of a specimen in the Whitby Museum, "No. 458, Robin Hood's Bay."

    s. Transverse section.

    s. Longitudinal section of a specimen in Prof Phillips's Collection. Robin Hood's Bay.

51. Belemnites scabrosus, n. s.

    l. Lateral view of a specimen in the Whitby Museum, "No. 976." From north side of Robin Hood's Bay.

d, v. Dorsal and ventral aspects.

52. P. *l*. Belemnites cylindricus. Lateral view of a specimen from Rosedale (near Staithes), in beds below the Jet-rock.

P. *v*, P. *d*. Ventral and dorsal view of the same.

52, M. *l*. Belemnites paxillosus.

    M. *l*. Ventral aspect of a specimen from the marlstone of Ilminster, in the Collection of Mr. Moore. (The figure should have been marked M. *v*.; the slight seeming ventral groove near the apex is an accidental effect of the shading.)

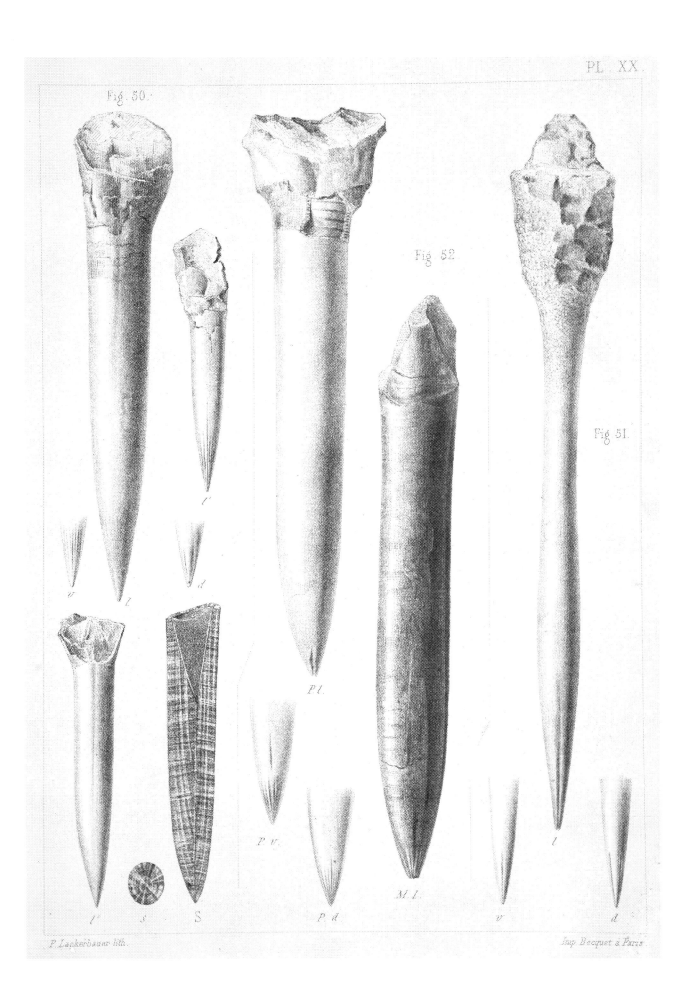

Fig. 50.

Fig. 52

Fig. 51

P. Lackerbauer lith.

Imp Becquet à Paris.

A MONOGRAPH

OF

BRITISH BELEMNITIDÆ.

BY

JOHN PHILLIPS,

M.A. OXON., LL.D. DUBLIN, LL.D. CAMBRIDGE, D.C.L. OXON., F.R.S., F.G.S., ETC.,
PROFESSOR OF GEOLOGY IN THE UNIVERSITY OF OXFORD.

PART IV,

CONTAINING

PAGES 87—108; PLATES XXI—XXVII.

LONDON:

PRINTED FOR THE PALÆONTOGRAPHICAL SOCIETY.

1869.

PRINTED BY
J. E. ADLARD, BARTHOLOMEW CLOSE.

Proportions. The diameter (*v d*) being taken at 100, the ventral radius is 40, the dorsal 60, the cross diameter 100, the axis 420.

PHRAGMOCONE. Nearly straight, with a nearly circular section; the angle $m = 25°$.

Locality. In the lower part of the Upper Lias at Saltwick (*Phillips*); at Robin Hood's Bay (*Cullen*); in ironstone layers at Kettleness (*Simpson*); in the shale under the Jet-bed, plentifully; and in ironstone layers at Staithes and Rosedale (*Phillips*); in the Marlstone series below the ironstone.

Observations. The agreement of this Belemnite with that long known as *B. paxillosus* is obvious and intimate, and the resemblance of particular selected specimens is almost complete, the principal observable difference being a greater proportionate length of axis and a longer tapering to a less obtuse apex in the Yorkshire specimens.

For comparison, a specimen from Ilminster, in Mr. Moore's Cabinet, is represented fig. 52, *Ml.*

Recurvation of the apex occurs in several of the specimens of *B. paxillosus*, especially in those from Ilminster; in several of the specimens of *B. cylindricus* from Rosedale, near Staithes, it is so pronounced as to approach the form of *B. aduncatus.*

On the whole, I can hardly doubt that the Yorkshire specimens agree with *B. paxillosus amalthei* of Quenstedt ('Cephal.,' pl. xxv, fig. 5); the state of conservation seldom allows of the striation of the apex to be perfectly seen, as in our representation of *B. paxillosus* (Pl. XX, fig. 52, *Ml*).

B. elongatus, B. apicicurvatus, B. paxillosus, and *B. cylindricus,* taken together, compose a natural group of generally cylindrical or cylindroid forms, with dorso-lateral grooves at the apex, and plaits or striæ on the ventral and dorsal aspects (exceptionally, a deeper stria on the ventral and also on the dorsal face). They are unknown in Lower Lias, but extend from the base of the Middle Lias to the lower part of the Upper Lias, and are found in Dorsetshire, Gloucestershire, Northamptonshire, Lincolnshire, and Yorkshire.

BELEMNITES OXYCONUS, *Quenstedt.* (Diagram, No. 23, p. 88.)

Reference. *Belemnites (tripartitus) oxyconus,* Quenstedt, 'Cephalop.,' p. 419, pl. xxvi, figs. 19, 20, 1849.

GUARD. Compressed, conoidal or conical, ending in a produced, pointed, somewhat reclined apex; lateral grooves extend over the alveolar region.

Transverse section oval, the ventral region broadest.

Locality. Cheltenham, in the Belemnite-bed of the Lower Lias (*Buckman*).

The fossil here represented, from Mr. Buckman's Collection, is from the "Belemnite-bed" of the Lower Lias of Cheltenham.   It corresponds to all appearance with the species

DIAGRAM 23.

Belemnites oxyconus.

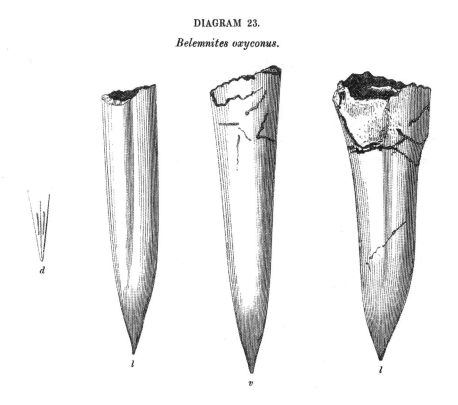

referred to by Quenstedt, and I find it also to be much allied to *B. elegans* of Simpson (p. 84 ; and Pl. XX, fig. 50) ; but it is not striated about the point, as that species seems always to be ; its axis is shorter, and the figure is more oblique.   It seems allied to *B. acutus* of Miller, and may be the old form of that species.

DIAGRAM 24.

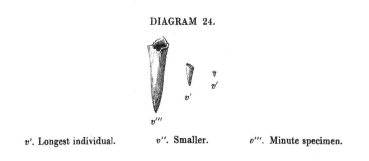

v'. Longest individual.   *v''.* Smaller.   *v'''.* Minute specimen.

In Mr. Charles Moore's very interesting collection from the Upper Lias of Ilminster are specimens of an extremely small, short Belemnite, which may possibly be the young of such a species as *B. Voltzii.* It merits, however, a careful description, as all the very young Belemnites do.

B<small>ELEMNITES</small> <small>MINUTUS</small>, n. s. (Diagram, No. 24, *v′, v″, v‴.*)

G<small>UARD</small>. Short conical, straight, acute, with two faint latero-dorsal grooves, and an obscure trace of ventral depression towards the apex.

Dimensions. From ⅛ to ¼ and (apparently the same species) ¾ inch long.

Locality. Upper Lias, Ilminster.

Remarks. It is possible that, by further research, older forms may be identified with these almost microscopical specimens of Mr. Moore's acuteness and industry. They cannot be the young of *B. Ilminstrensis*, but rather should be referred to a form such as *B. Voltzii.*

In the same collection are specimens from the Upper Lias of Ilminster of a slender Belemnite, agreeing in general with *B. quadricanaliculatus*, but differing in the striation and some other points of interest, as will appear by the following description, in which it is regarded as a variety. Specimens of the ordinary type occur with it.*

B<small>ELEMNITES</small> <small>QUADRICANALICULATUS</small>, var. <small>OBSOLETUS</small>. (Diagram, No. 25, *v′, v″, v‴, l, d.*)

G<small>UARD</small>. Long slender conical, often a little bent, marked by a conspicuous double-striated ventral groove, extending over part of the alveolar region; striæ and traces of dorsal grooves near the apex, which is always found truncated.

Dimensions. Under 1 inch in length.

Locality. Upper Lias, Ilminster.

DIAGRAM 25.

v′. Ventral face, full size.   *v″.* The same, smaller.   *v‴.* Magnified view of the ventral striæ.   *l.* Lateral, *d.* dorsal.

---

* In the description of *B. quadricanaliculatus*, p. 68, the dorsal groove is said to be sometimes double. I find, also, the ventral groove double in several examples.

In the preceding pages I have endeavoured to describe all the forms of Liassic Belemnites of which I have been enabled to study sufficient examples. Under very favorable circumstances a few characteristic forms have been traced completely from very young to quite mature specimens—for example, *B. Ilminstrensis* (Pl. XII). Some have been collected in such great numbers as to leave no doubt of their possessing a real relationship—*e. g. B. lævis* (Pl. X); others possess such peculiarities of form and structure, amid many variations, as to allow of their being quoted without fear from very distant localities—*e. g. B. clavatus* (Pl. III).

Some cases have occurred of forms which, if not really very rare, have not yet been collected in sufficient number to allow of giving more than the description fitting to a particular age of the animal, as *B. excavatus, B. calcar, B. dens,* on Pl. II. In handling these curious species I have been guided by the experience acquired in examining the other more favorable cases, and I hope the results will be found trustworthy in the main. But I am most anxious to be furnished with more, many more, examples of Belemnites of various magnitudes, from the largest to the least, which may be collected from *one limited band of Lias,* at one definite place. For instance, in the greatly reduced band of Upper Lias clays in the Vale of Gloucester, near Dursley and Wotton-under-Edge, Belemnites occur with *Ammonites bifrons,* as well as others in the Marlstone below, and in the sands and Cephalopodal bed above. If any one to whom this may reach would devote a few days to extract a few dozens of Belemnites of all magnitudes from each of the beds named, taken separately, preserving carefully the fragments which belong to each specimen, and would send them to me for study, the conclusions to which I shall soon request attention as to the geological succession of Liassic life, and the changes of form which accompany the transition from Liassic to Oolitic sediments, might acquire a considerable augmentation of value. I did once make such a collection (1843), and was in hopes that it might have been preserved in a museum which contains very many palæontological data gathered under my direction, but ill-fortune befell them after they left my hands.

The reader will doubtless remark that of the Liassic species mentioned by previous English authors the following are not described by me:

> *B. aduncatus* of Miller, said to be from Weymouth and Lyme, pp. 5 and 8.
> *B. trifidus,* Voltz, Whitby, p. 8.
> *B. brevirostris,* D'Orbigny, Cheltenham, p. 11.

Of the first I possess a specimen—Liassic. It appears to me an exaggerated variety of *B. elongatus.*

Many Liassic Belemnites exhibit a tendency to recurvation of the apex, the ventral portion being more or less curved upwards or towards the back. Even when both the ventral and dorsal outlines curve to meet, or towards meeting, the ventral curve is often most decided, and carries off the apex towards the dorsal line, so as to make it project in that direction. In old specimens of *B. vulgaris* at Whitby, *B. cylindricus* at Staithes, and *B. paxillosus* at Ilminster, this is very obvious. Thus, in a series of twenty-four specimens of *B. cylindricus* collected by myself near Staithes, about half are positively recurved, the rest slightly so or not at all. When the recurvation is very decided it is often accompanied by ventro-planation. Two specimens of *B. elongatus* from Cheltenham in my collection, which were presented to me by my friend the late H. E. Strickland, Esq., show this variation strongly; it seems to be not at all an anomaly, but the usual structure. The recurvation, where regular, increases with age, and thus augments the tendency to bluntness which appears in many old specimens.

The variation here alluded to under the title of recurvation is so far normal that it is always an upward curve.

The one specimen on which *B. trifidus* was quoted was from Upper Lias, Whitby. It is not distinct from *B. tripartitus*. Of *B. brevirostris*, D'Orb., quoted by Professor Morris from Cheltenham, I have seen no specimens.

I am aware that in the collections preserved at Whitby several interesting specimens occur to which Mr. Simpson has assigned names in his treatise on the Yorkshire Lias, and I have made drawings and careful observations of many of them; but the difficulty, to which reference has been made above, of deciding upon claims to specific distinction from single, perhaps exceptional, specimens, deters me from doing more at present than expressing my hope that the diligent Curator of the Whitby Museum will persevere in the useful labour of amassing very many examples of the species which occur in each bed of the Lias which he knows so well. For thus, I believe, he will have just grounds for reducing the number of his specific names, and of augmenting the notices of variety under each form.

One of these interesting fossils I must, however, mention expressly, as it occurs in the Marlstone of the coast very frequently, but seldom in so good a condition as to admit of sufficient definition. It is that called *Belemnites inæqualis* by Mr. Simpson, belonging to the group of bisulcate Belemnites allied to *B. apicicurvatus, paxillosus*, &c.. It appears to be identical with specimens of *B. paxillosus numismalis*, of Quenstedt ('Cephalop.,' t. 23, fig. 21), which I have lately examined at Stutgardt. It is included in the Table, p. 94.

Oppel, in his work entitled 'Jura-Formation' (1856—1859), mentions *B. Whitbiensis*, from the lower part of the Upper Lias. I am convinced this must be a synonym of *B. vulgaris*, one of the forms of Belemnites from this zone already named and described by Mr. Simpson in 1855. This zone is called by Mayer, Toarcian I, *a—c*.

He also names a species from Bridport, as accompanying *Ammonites torulosus* in the lowest part of the Inferior Oolite, *B. Dorsetensis*. From inspection of the Munich Collection I infer this to be a synonym of *B. quadricanaliculatus*. The beds thus designated are included by Mayer among the Upper Lias beds, and called Toarcian III.

Mayer, in his "Systematic List of Jurassic Belemnites" ('Journal de Conchyliologie,'

1863), quotes Oppel for a species named *B. Wrightii* after the eminent palæontologist of Cheltenham, and assigns it to the upper part of the Upper Lias, viz. Toarcian I, *d.* I have not been fortunate enough to obtain a specimen. He also gives a new name (*B. neglectus*) to a specimen said by Oppel to be from the Belemnite bed of Lyme Regis, which is figured by (Quenstedt, 'Jura,' pl. xlii, fig. 20) as from the lowest bed of Inferior Oolite. If it be "*unicanaliculate*," as Mayer says, it is probably of an Oolitic type, or from the Bridport Sands.

It will be requisite, no doubt, hereafter to construct a Supplement to these pages, and my own collection contains a few specimens worthy of notice if more, and more illustrative, specimens, can be obtained.

Meantime it appears to me useful to present a summary of the distribution of well-recognised forms in the Yorkshire Lias, every bed of which between Saltburn and Saltwick I have in late years carefully examined and for the most part exactly measured, and searched many times for Belemnites and other fossils. The general result of this labour appears in the Tabular View,* p. 93, which is reduced from a larger drawing, beginning with Gryphitic beds above the Lima beds and *Ammonites Bucklandi,* and ending below the Dogger, and is not meant to include the transition sandy beds, if they may be so termed, at Blue Wick.

My friend Mr. Simpson has already composed a similar and more minute description of the beds between Saltwick and the Peak, so that the two may be compared, and the reference of every fossil on the Lias Cliffs of the Yorkshire coast to its real repository become by degrees quite complete. Far clearer on this magnificent coast than even on that of Dorset is the distribution of the fossils in the Middle and Upper Lias, and capable of often-repeated proof; but in respect of the lowest Lias, it is not in the Yorkshire cliffs that it must be studied. Nor do I know of more than one example of Rhætic beds in the county, and that is now concealed under the grassy surface of a deep cutting at Barton, on the North-Eastern Railway. It was exposed, many years since, in an anticlinal resting on the Keuper Marls, and I examined it well, but found few fossils in it.

* In the Table, p. 93, nodules are marked by interrupted bands. Two Belemnite-beds are marked by 'β. β.'

LIAS OF THE YORKSHIRE COAST BETWEEN SALTWICK AND SALTBURN.

Leda ovum.	Belemnites	Voltzii.	
	Belemnites	subaduncatus.	Amm. bifrons.
50			
	Belemnites	β. β. striolatus.	Amm. communis.
	Belemnites	β. β. subtenuis.	
100			
	Belemnites	dorsalis.	
	Belemnites	tubularis.	Amm Mulgravius (serpenti-nus).
	Belemnites	lævis.	
	Belemnites	tripartitus.	
150			
Posidonomya.	Belemnites	cylindricus.	Amm. annulatus.
200			
	Belemnites	acuminatus.	
Pecten æquivalvis.	Belemnites	paxillosus.	Amm. Hawskerensis (spinatus).
	Belemnites	breviformis.	
Avicula cygnipes.	Belemnites	rudis.	
250			
300		numismalis	Amm. Clevelandicus (margari-taceus).
Gryphæa depressa.			
350			
Avicula inæquivalvis.			
Modiola scalprum.	Belemnites	pollex.	Amm. maculatus.
400			
450			
Pholadomya ambigua.	Belemnites	elegans.	
	Belemnites	clavatus.	
500			
550			
600			
	Belemnites	dens.	
Gryphæa Maccullochii.	Belemnites	acutus.	Amm. obtusus.
650 feet			

Belemnites vulgaris.

Cardium truncatum.

In the Table of Distribution of Belemnites given at p. 93, I have introduced the names of several well-known and definitely characterised Ammonites in their ordinary places in the strata. This might have been done to a greater extent, and thus a more minute comparison have been made with the similar and equally well-known Ammonitic zones on the Dorsetshire coast. But this subject will come under a more full examination hereafter, when the Belemnitic beds above that which is so famous at the foot of Golden Cliff shall have been as fully explored as have been the contemporaneous parts of the Yorkshire section. Mr. Day and Mr. Etheridge have already made good progress in this work, which is not so easy on the Dorsetshire as on the northern coast.

I now present, in a diagram, a classification of Yorkshire Lias, according to natural divisions of the series, adopting as few groups as possible; and in this scale of time I have written the species of Belemnites already described from the cliffs of Saltburn, Staithes, Whitby, and Robin Hood's Bay.

Distribution of certain species of Belemnites in the strata of the Yorkshire coast.	B. acutus.	B. dens.	B. penicillatus.	B. elegans.	B. scabrosus.	B. pollex.	B. inæqualis.	B. rudis.	B. paxillosus. cylindricus.	B. acuminatus.	B. breviformis.	B. tripartitus.	B. lævis.	B. tubularis.	B. striolatus.	B. subtenuis.	B. dorsalis.	B. vulgaris.	B. Voltzii.	B. subaduncatus.
Alum Shale. Leda Beds	...	...	...	...	...	...	...	...	...	...	...	...	...	...	...	...	...	*	*	*
Upper Ironstones, &c.	...	...	...	...	...	...	...	...	...	...	...	...	*	*	*	*	*	*	...	...
Jet Series	...	...	...	...	...	...	...	...	*	...	...	*	...	...	...	...	...	...	...	...
Mid. Ironstone Series	...	...	...	...	...	...	...	*	*	*	*	...	...	...	...	...	...	...	...	...
Marlstone	...	...	...	...	...	...	*	...	...	...	...	...	...	...	...	...	...	...	...	...
Lower Ironstone Series	...	...	...	...	...	*	...	...	...	...	...	...	...	...	...	...	...	...	...	...
Lower Lias Shale	*	*	*	*	*	...	...	...	...	...	...	...	...	...	...	...	...	...	...	...
Lias Limestones	...	...	...	...	...	...	...	...	...	...	...	...	...	...	...	...	...	...	...	...
Rhætic beds	...	...	...	...	...	...	...	...	...	...	...	...	...	...	...	...	...	...	...	...

LIAS—OOLITE.

A complete passage by continuous change from the Upper Lias to the Inferior Oolite, from an argillaceous shale or clay to an Oolitic limestone, is not to be looked for ; but there are a few localities in England where beds are interposed which mark one or more stages in the change of sediments, and are themselves marked as much by change of the forms of life as by mineral variation. My attention was attracted to this subject during frequent examinations of the Yorkshire coast on several occasions previous to the publication of the first volume of my work on the strata and fossils of that coast, in 1829. Generally speaking, the Oolitic series terminates below in a variable sandy, irony, or calcareous rock, sometimes almost full of shells, at other places not yielding one. The Lias on which this rests is in general strongly contrasted with it in colour, structure, composition, and fossils ; but in one locality on the coast, at Blue Wick, under the Peak of Robin Hood's Bay, the series of strata, including Lias below and Oolitic rocks above, admits of subdivisions which soften the change from Lias to Oolite, and exhibit a pretty full series of fossils for illustration of the life forms of the transition period.

In my account of these beds (' Illust. of Geol. Yorkshire,' vol. i, p. 91, first edition) they are all classed as a conchiferous (*Dogger*) series analogous to the Inferior Oolite of Bath, which at that time was universally allowed to include the sandy beds below, so well known and described at Bath and Yeovil. The description given of these beds at Blue Wick shows that they were regarded by me as " gradually changing in the lower beds" to the Alum-shale.

In 1859 the further researches of Dr. Wright (' Geol. Journal,' vol. xvi, p. 1) convinced him that these passage-beds were the equivalent of the " Cephalopodal bed" and sands which cap the Lias of Gloucestershire and Dorsetshire ; and by the valuable evidence of the Ammonitidæ they have been of late years pretty generally associated with the Lias, as the uppermost member of that formation. Of the few Belemnites which occur in these beds I have noticed the most conspicuous, viz. *B. irregularis* (Pl. XV, fig. 37), at Frocester Hill ; *B. inornatus* (Pl. XVIII, fig. 46), at Blue Wick ; *B. sulci-stylus*, (Pl. XIX, fig. 49), at Nailsworth.

Having lately on several occasions examined many times with great care these sands at Bridport, where they are capped by the Oolite, and again at Yeovil under similar conditions, I am able to add to the list of species four other forms, viz.—

Bel. Voltzii. Upper part of the Yeovil Sands, only 15 or 20 feet below the Oolite.
Bel. tricanaliculatus, Quenstedt, ' Cephal.,' t. 25, figs. 13—15. Bridport Sands.
Bel. quadricanaliculatus, Quenstedt, ' Jura.,' t. xli, f. 17.

14

Bel. unisulcatus, Blainville, 'Mém. sur les Bel.,' pl. v, fig. 21; D'Orbigny, 'Ter. Jur.,' pl. viii, figs. 1—5.　Bridport and Bradford Abbas.

I regard these as all truly Liassic forms, to whatever extent they may hereafter be found continued into the Oolitic strata.

BELEMNITES OF THE OOLITIC SYSTEM.

Regarded from the most general point of view, the Belemnites of the Oolitic lime-stones, sands, and clays present themselves in five natural groups, which may be thus typified :

1. Group of *Belemnites giganteus*, Blainville.　Large compressed species, with a nearly regular elliptical or oval section ; no ventral groove.

2. Group of *Belemnites canaliculatus*, Schlotheim (*sulcatus*, Miller).　More or less depressed ; the ventral surface conspicuously grooved in the middle part of the guard.

3. Group of *Belemnites hastatus*, Blainville, whose remarkable elongation, hastate shape, and deep ventral groove, mark them distinctly.

4. Group of *Belemnites tornatilis*, Phillips (*Owenii*, Pratt ; *Pusozianus*, D'Orbigny). Long subcylindrical Belemnites, with a groove on the ventral aspect towards the apex of the guard.

5. Group of *Belemnites abbreviatus*, Miller (*excentricus*, Blainville).　Large short Long Belemnites, plane or broadly grooved on the sides, flattened or slightly grooved near the apex on the ventral aspect.

I propose to describe these groups in the order set down ; and have only now to remark, by way of introduction, that the first group may be regarded as continued from the Upper Lias into the Bath Oolite series, where apparently it grows to the utmost magnitude, and then ends.　The second group begins in the lowest of the Oolites and ascends to the Oxford Clay, not, I believe, to the uppermost part of that deposit.　The third group, notwithstanding its seeming resemblance to *Belemnites clavatus* of the Lias, is really more allied to the second here noted ; it begins in the Bath Oolite series, but not, I believe, at the base of it, and extends into the Kimmeridge Clay.　Its relation to *B. jaculum* of the Speeton Clay and *B. pistilliformis* of the Neocomian beds, and the small Belemnites of the Folkestone Gault, will be considered hereafter.　The fourth group extends from the Kelloway Rock to the Kimmeridge Clay.　The fifth begins in the Oxford Clay, not, I believe, at the base, and extends upwards into the Kimmeridge Clay of Oxfordshire, the Speeton Clay of Yorkshire, and the congeneric bed of Lincolnshire called Tealby Stone.

On the Group of large compressed Belemnites in the Inferior Oolite (Pls. XXI to XXIV).

Belemnites ellipticus (Pl. XXI) is the name given by Miller to a fine, straight, compressed species, which occurs in the Oolite of Dundry, and in the country near Yeovil and Bridport.

Sowerby (in 'Min. Conch.,' pl. 590, fig. 4) figures under the name of *B. compressus* of Blainville one of the several forms of large Belemnites which are frequent in the " Grey Limestone," a part of the Bath Oolite series, of Gristhorpe and other places near Scarborough. Some of these fossils agree exactly with *B. quinquesulcatus* of Blainville as to the termination, others correspond with *B. Aalensis* of Voltz in general figure, while examples may be selected which seem to be identical with *B. gladius* of Blainville and *B. giganteus* of Schlotheim. They do not occur in the strata above.

In the south of England such forms are not frequent. I have, however, been favoured with the sight of two fine examples from the collection of Mr. Read, of Salisbury, obtained from near Sherborne; and another has been sent me from Leckhampton by Mr. Buckman.

In considering how to deal with these fossils, I remark, in the first place, that *B. ellipticus* of Miller, from Dundry, is not exactly to be matched in form and structure by any specimens from Yorkshire, unless a single specimen from the grey Dogger-beds of Blue Wick be referred to it. Next, that in specimens from Yorkshire two distinguishable variations appear—1, analogous to *B. giganteus*, *B. gladius*, and *B. Aalensis*; 2, analogous to *B. quinquesulcatus* and *B. compressus* of Blainville. And these same forms occur in the south of England, so that we have three species or remarkable varieties in this group of large Belemnites to be considered. Those who regard them as varieties will still find it useful to preserve the distinctive names, though all may be spoken of as *Belemnites giganteus*, Auctorum. I begin with *Belemnites ellipticus* of Miller.

BELEMNITES ELLIPTICUS, *Miller*. Pl. XXI, fig. 53.

Reference. *B. ellipticus*, Miller, 'Geol. Trans.,' 2nd series, vol. ii, p. 60, pl. viii, figs. 14—16, 1826.

GUARD. Straight, elongate, very much compressed, gradually and uniformly tapering, with an almost uniformly oval section (the ventral face widest), smooth, without furrows except near the summit, where two or more faint lateral facettes break the regularity of the surface.

The transverse sections of the sheath are almost uniformly oval, the ventral face being rather wider than the dorsal, and the axes measuring 72 and 100. At the apex of the alveolus the inner layers have a somewhat less oval figure than the exterior; and very near the axis they become undulated by the latero-dorsal facettes, but never show the deep grooving of *B. Aalensis, B. quinquesulcatus,* &c. Miller says the inner layers give a nearly circular section, but this is not the case in either of my specimens. Substance light-coloured, finely fibrous.

Greatest length of axis 7½ inches in a specimen 12½ inches long; the diameters at the apex of the alveolus being under 1 inch from back to front, and above ⅝ths from side to side.

Proportions. The diameter, from front to back, at the apex of the alveolus being 100, that from side to side is 72; the ventral radius 36, the dorsal 64.

PHRAGMOCONE. Incompletely known, but presenting in fragments remarkable characters; oblique, with straight sides inclined at angles of 16° and 20° nearly. Section compressed elliptical, with axes as 88 to 100 near the apex, and as 83 to 100 near the aperture, the excentricity increasing with age. The septa are oblique, with slightly waved edges, siphon oval, submarginal, in a slight degree removed from the conjugate axis towards the right side of the animal.

Depth of the largest chamber = one seventh of the diameter.

Locality. In the Inferior Oolite of Dundry Hill. *Miller's* Collection in the Bristol Museum. In *Mr. W. Sanders's* Collection and *Prof. Phillips'* Collection.

Observations. Miller's figures are very unsatisfactory, and his reference of fig. 17 in his essay to this species is a mistake. Many additional specimens are needed, especially young examples, but I think the main characters are clear for adults. The surface of some specimens is much worn and eroded; Serpulæ adhere to others. *Belemnites longus* of Voltz (Pl. III, fig. 1) is similar, but its apicial line is uniformly less excentric, and shorter in proportion; it is from the Oolite of Buxweiler. *Belemnites gladius* of Blainville, from the Oolite of Bayeux, and from Rabenstein, is analogous if not identical.

BELEMNITES AALENSIS, *Voltz.* Pl. XXII, fig. 54; Pl. XXIII, fig. 55.

Reference. *Belemnites giganteus,* Schlotheim, 'Petref.,' p. 45 (probably), 1820.
B. *gigas,* Blainville, 'Bélemn.,' p. 91, pl. v, fig. 20 (probably), 1827.
B. *Aalensis,*Voltz, 'Obs. sur les Bélemn.,' p. 60, pl. iv, pl. vii, fig. 1, 1830.
B. *Aalensis,* Phillips, 'Geol. of Yorkshire,' p. 166, 1835.

GUARD. Sheath large, very much compressed, smooth. Anterior region unknown. Alveolar region cylindroid, often a little contracted in the middle. Apicial region (frequently) much contracted at a small distance from the alveolus, and thence extending into a long, compressed (often slightly bent or undulated) conical figure, terminated obtusely, obscurely striated lengthwise, and marked by six or seven furrows, viz. two latero-dorsal, very deep, much longer than the others, extending nearly the whole length of the attenuated part of the apicial region; four latero-ventral, of unequal length, the shortest and faintest being near the ventral line; and one medio-dorsal, always faint, and sometimes absent.

The sections of the sheath vary according to the distance from the apex. The apicial line is straight. At the alveolar apex the external layer is oval, with diameters as 81 to 100, the dorsal part being widest; nearer the apex the figure is formed by two unequal, nearly semicircular curves, the ventral one being largest; at the apex it is extremely compressed, with six or seven grooves. These grooves show in the central portions of every section of the sheath, and on breaking the specimens across a central prominence occasionally appears, but less distinctly than in the next species. The substance is compact and light-coloured, breaking nearly at right angles to the axis.

PHRAGMOCONE. Oblique, with straight sides inclined at angles of 23°, the back and front inclined at 27°. Section elliptico-compressed, with diameters as 91 to 100.

Greatest length observed 20 inches, of which the apicial line is 12. Greatest diameter on the middle alveolar region 2 inches.

Proportions. The diameter, *v d*, at the alveolar apex being 100, the lateral diameter is 81, the ventral radius is 40, the dorsal 60.

VARIETIES. In most specimens the apicial region contracts remarkably at about one third of its length behind the alveolar apex, while before that line the alveolar region is nearly cylindroid; in others the whole figure more nearly approaches to a cone; the lateral profile of some is bent in a gentle arch, in others undulated.

Locality. In the Lower Oolite formation (upper part) of Yorkshire, especially on the Scars at White Nab, south of Scarborough. (At Aalen in Wurtemburg, in the Lower Oolite formation, lower part, *Voltz*). In *Mr. Bean's* Collection, Yorkshire Museum, Scarborough Museum, and in the *Author's* Cabinet. In the Lower Oolite of Sherborne (*Mr. Reed*); and in the Lower Oolite of Leckhampton (*Mr. Buckman*).

Observations. On specimens from White Nab, thin Oyster shells are found attached, and accommodated to the curved surface of the alveolar region of the sheath. In the alveolar cavity of another are crystals of sulphide of zinc.

BELEMNITES QUINQUESULCATUS. Pl. XXIII, fig. 56; Pl. XXIV, fig. 57.

Reference. *Belemnites quinquesulcatus*, Blainville, ' Mém. sur Bélemn.,' p. 83, pl. ii, fig. 8, 1827.

B. compressus, Sowerby, ' Min. Conch.,' pl. 590, fig. 4, 1828.

GUARD. Sheath large, conical, compressed, smooth. Anterior region unknown. Alveolar region conoid. Apicial region rapidly tapering to an obtuse compressed summit, and grooved with four, five, or six furrows, viz. two latero-dorsal extending over about one third of the apicial region, two latero-ventral of about half that length, one medio-dorsal, and one medio-ventral of variable length and distinctness.

The sections of the sheath vary according to situation. At the alveolar apex the contour is nearly elliptical, with diameters as 90 to 100; within the border the layers are oval (the ventral region broadest), while nearer the centre the layers show inflexions corresponding to the grooves of the summit, though still nearer the axis these are again lost. Nearer the summit the contour consists of two unequal subcircular curves, and the plan of the summit is a compressed pentagonal or hexagonal figure, with four, five, or six deep notches. When the sheath is broken across the central layers sometimes separate from the rest, so as to appear in a fluted prominence, and the same thing happens in old specimens from decay. Substance light-brown or grey.

Greatest length observed 8 inches, with a diameter not exceeding 2 inches.

Proportions. The ventro-dorsal diameter at the alveolar apex being 100, the ventral radius is 43, the dorsal 57; the transverse diameter is 87.

PHRAGMOCONE. Oblique, deeply inserted, with straight sides inclined at angles of 23° and 27°. Section compressed elliptical in the ratio of 87 to 100. The ventral region occupies about half the circumference. The septa are very oblique, advancing dorsally, not waved on the edge nor undulated by the siphon. Irregular striæ are seen parallel to the ventral line, crossing others parallel to the edges of the septa. On the dorsal aspect are arched shades of colour crossing the contrary curvatures of the striæ of growth. The ventral aspect is often black.

Proportions. Diameters as 100 to 87. Depth of chambers one eighth of the diameter.

VARIETIES. The degree in which the figure approaches to a cone varies a little, and the apical furrows are rather inconstant. Some specimens show hardly any medio-ventral or medio-dorsal grooves, in others these are distinct or even duplicate, and the intervening spaces are striated.

Localities. In the Lower Oolite formation of Yorkshire, at White Nab, Cloughton Wyke, Carlton Husthwaite. (Also near Mezières, *Blainville.*) At Sherborne, in Dorset

(*Mr. Read*).   Specimens occur in the Yorkshire Museum, Museums of Whitby, the Scarborough Phil. Soc., &c.

Observations.   Oyster shells adhere frequently to the apicial region of the sheath, which is always more or less incomplete and eroded.   This short form is conjectured by D'Orbigny to be the female of *B. giganteus* (*B. Aalensis,* Voltz).   The two forms must be regarded as closely allied.

On the Canaliculated Belemnites of the Inferior Oolite (Pl. XXV).

Miller, in his account of *Belemnites sulcatus* (' Geol. Trans.,' 2nd ser., vol. ii, pl. viii figs. 3, 4, 5), gives for localities, " Dundry, near Oxford, Inferior Oolite."   His fig. 3 appears to represent *B. apiciconus* of Blainville, which occurs frequently in the Inferior Oolite, but has not yet been found near Oxford.   Fig. 5 I have always supposed to represent a fossil from the Oxford Clay.   It seems to be copied or modelled from specimens which still exist in the Bristol Museum, and are marked " *B. sulcatus,*" Inferior Oolite.

In the lowest beds of the Inferior Oolite of the south of England, generally, among the most frequent Belemnites are those of the type of *B. apiciconus,* Blainville.   To judge from examples collected by Mr. Buckman near Sherborne, and by myself near Yeovil, there are two other distinguishable forms, of a slenderer figure, one canaliculated to the apex or very near it, the other not carrying its groove so far backward.   To these there may be added the fossil called in my work on the Yorkshire Coast *B. anomalus,* and there quoted from the Kelloways Rock.   It belongs really, I believe, to the Grey Limestone of Gristhorpe.   None of these, so far as I yet know, have the fusiform or hastate shape in any period of their growth; but I have not met with very young forms of any one of them.   They are all distinct from the canaliculated Belemnites of Stonesfield, and never exhibit much of that depression in the post-alveolar region which always belongs to the allies of *B. fusiformis* of Miller.

Belemnites apiciconus, *Blainville.*   Pl. XXV, fig. 58.

> *Reference.*   *Belemnites apiciconus,* Blainv., ' Mém. sur les Bélemnites,' p. 69, pl. ii,
>   fig. 2, 1827.
>   *B. canaliculatus,* Quenstedt, ' Der Jura,' p. 411, pl. lvi, fig. 6, 1858 ;
>   ' Cephalop.,' p. 439, pl. xxix, fig. 6, 1849.

Guard.   Cylindrical in the middle, tapering in a curve to a pointed apex.   Ventral surface marked by a deep narrow groove, which is continuous for the whole length, except toward the apex, where a portion in length about equal to half the axis is free from any groove or depression.

Transverse section nearly circular; axis subcentral.

Dimensions of an ordinary specimen $2\frac{1}{2}$ inches, of which $1\frac{1}{2}$ belong to the axis. Diameter at apex of alveolus 0·46.

Proportion of axis to diameter at apex of alveolus 325 to 100.

PHRAGMOCONE.   Unknown to me.

Locality.   Yeovil, in Inferior Oolite (*Oxford Museum*).

Observations.   D'Orbigny quotes *Bel. sulcatus* as the prior equivalent of this species, and gives figures ('Terr. Jur.,' pl. xii, figs. 1—8) of the young and old, with several cross sections, and one longitudinal section.   It is very doubtful whether these all belong to one species.   The fusiform young is marked by an almost continuous furrow, like the Stonesfield fossils; the longitudinal section belongs to a remarkably short type; the mature individual has a peculiarity in the expanding posterior ending of the canal.

In the Museum of the Garden of Plants is a large series of '*B. apiciconus*,' jun. from Croiselles, in Normandy.

Bel. canaliculatus of Schlotheim is too variously interpreted to be safely quoted, except in the definite shape given to it by Quenstedt, in the work quoted.   In his 'Cephalopoda,' *B. canaliculatus* is made to include *B. sulcatus* of Miller, *B. Altdorfensis* and its Russian analogue, and the Stonesfield fossil, of which a young example is given, 'Cephalopoda,' pl. xxix, fig. 7.

BELEMNITES BLAINVILLII, *Voltz.*   Pl. XXV, figs. 59, 60.

Reference.   *Belemnites acutus*, Blainville, 'Mém. sur les Bélem.,' p. 69, pl. ii, fig. 3 (medium size), 1827.

     B. Blainvillii, Voltz, 'Obs. Bélemn.,' p. 37, pl. i, fig. 9 (full-sized), 1830.

      ,,   D'Orb., 'Terr. Jur.,' p. 107, pl. xii, figs. 9—16 (young), 1842.

GUARD.   Elongate, uniformly tapering to a smooth, rather blunt apex.   Ventral face marked by a distinct narrow groove extending from very near the apex to the beginning or over a part of the alveolar cavity, and then ceasing gradually.

Transverse sections nearly circular, or a little oblong, with a nearly central axis.

Dimensions.   Largest specimen which I have measured $4\frac{1}{2}$ inches long, of which the axis is $3\frac{1}{4}$ inches, the diameter at the alveolar apex being 0·43 from back to front and 0·41 across.   In smaller specimens length 3 inches, diameter 0·36.   Young specimens, such as figured by Voltz, I have not seen.

Proportions. The axis of the guard measures 600 to 750, the diameter being 100.

PHRAGMOCONE. Indistinctly seen.

Locality. Inferior Oolite, at Bradford Abbas (*Buckman*), Sherborne (*Oxford Museum*).

BELEMNITES CANALICULATUS, *Schlotheim.* Pl. XXV, fig. 61.

Reference. Belemnites canaliculatus, Schlotheim, 'Petref.,' p. 49, No. 7, 1820.

           ,,   D'Orb., 'Terr. Jur.,' p. 108, pl. xiii, figs. 1—5, 1842.

GUARD. Cylindro-conical, tapering uniformly to a somewhat sharp point. Ventral surface marked by a narrow deep groove along the whole length, so that even the point is hardly free from groove.

Transverse section nearly circular, in var. a through the whole length, in var. β depressed in the post-alveolar region.

Dimensions. Large specimen 2 inches and $\frac{3}{10}$ths, of which the axis occupies $1\frac{1}{9}$ inch, the diameter at the apex of alveolus being 0·4.

Proportions. Axis 300, the diameter being 100.

Locality. This fossil occurs in the lowest beds of Inferior Oolite at Bridport and Yeovil (*Phillips*), at Dundry (*Sanders*), Wotton-under-Edge (*Phillips*), and Leckhampton Hill (*Buckman*). How much further north it is to be found I cannot say. It has not yet been seen at Stonesfield, or in the Oolites north of Oxford (which are mostly of the Great Oolite), nor have I seen it in Yorkshire.

Observations. The specimens represented in Pl. XXV agree with the figure and description of D'Orbigny, except that they are not at all depressed in the post-alveolar region. Thus two marked varieties arise. They are much shorter in proportion than *B. Blainvillii,* and by the continuity of the ventral sulcus are easily separated from *B. apiciconus.*

BELEMNITES TERMINALIS, n. s. Pl. XXV, fig. 62.

GUARD. Elongate, lanceolate, or very slightly subhastate, depressed, tapering to an acute point; ventral face grooved over the alveolar region, and over the post-alveolar tract to near the apex; groove deep, narrow.

Transverse section at the alveolar apex wider than long; still wider in proportion towards the apex; axis nearer to the ventral face.

Proportion of axis of guard to the ventro-dorsal diameter at the alveolar apex 550 to 100.

PHRAGMOCONE. Unknown.

Locality. Yeovil, Inferior Oolite; specimen in the Oxford Museum.

Observations. This fossil differs from *B. apiciconus* chiefly by its general depression and smaller diameter when specimens of the same length are compared. It agrees in general figure with *B. Bessinus*, as given by D'Orbigny, and as represented at Stonesfield, but the groove does not reach so far backward. It may be supposed to be the connecting link between the two species named.

BELEMNITES ANOMALUS, *Phillips.* (Diagram No. 26.)

DIAGRAM 26.

Reference. *Belemnites anomalus*, Phillips. 'Geology of Yorkshire,' vol. i, p. 166 (2nd edition, 1835).

GUARD. Elongate cylindrical in the post-alveolar region, tapering to a long conical point. Ventral surface grooved in the alveolar and post-alveolar, but not in the apicial region, towards which the groove, previously narrow, grows shallower and wider.

Transverse section somewhat elliptical in the alveolar region, by reason of the ventral thickening of the guard; nearly circular in the post-alveolar region.

Greatest length observed 2·75 inches, of which the apicial line is 2·0; the greatest diameter 0·34.

Locality. White Nab, near Scarborough, in the Grey Limestone. Specimens in the Cabinets of *Mr. James Cook*, *Mr. Bean*, and the *Author*.

Observations. The greater proportionate length of the axis of the guard is the most marked differential character when this is compared with *B. apiciconus*.

ON THE BELEMNITES OF STONESFIELD, PL. XXVI.

Lhwyd, in the 'Lithophylacium Britannicum,' 1699, refers to "Stonesfield" for three specimens of Belemnites, all marked by a ventral furrow.

"1677. Belemnites major canaliculatus, sive aqualiculo per mediam longitudinem insignitus. È fodinis Stunsfieldiensibus.

"1705. Belemnites ari-pistillum referens, canaliculatus. È lapidicinâ Stunsfieldiensi " (figured in Lhwyd, tab. 25, fig. 1705, and copied in Diagram 27, below).

DIAGRAM 27.

"1720. Belemnites (cylindraceus) formæ compressioris, fissurâ altero latere donata, è fodinis Stunsfeldiensibus."

The specimens thus referred to by numbers cannot now be recognised in the Oxford ("Ashmolean") Museum.

Platt, in 1764, illustrates his sensible account of the 'Origin and Formation of the Extraneous Fossil called the Belemnite' by instructive figures of fusiform specimens from "Stonsfield" and Piddington, the latter being found in Oxford Clay. On his pl. iv, fig. 3 represents the Piddington fossil, with a groove not reaching to the alveolar region, while fig. 2 shows the completely grooved surface of the Stonesfield fossil, as may be seen by the diagram below, which is copied from Platt.

DIAGRAM 28.

Parkinson's 'Organic Remains of a Former World' presents to us the fusiform Belemnite of Stonesfield, in vol. iii, pl. 8, fig. 13 ; the lateral and retral expansion of the guard being greater than usual.

Miller, in the 'Geol. Trans.,' 2nd ser., vol. ii, pl. viii, fig. 22, gives a figure of the same unusual proportions, but with the peculiarity that the ventral groove does not extend over any part of the alveolar cavity, a circumstance never yet observed by me among upwards of twenty specimens from Stonesfield.

Morris and Lycett, in the 'Monographs of the Palæontographical Society,' give figures and descriptions of the Belemnites of Stonesfield, under the names of *B. Bessinus* (adopted from D'Orbigny) and *B. fusiformis* (vol. for 1850, pl. i, figs. 5 to 8). The figs. 5 and 7 for *B. Bessinus* differ scarcely in anything from figs. 6 and 8 for *B. fusiformis*, except that these latter are slightly swollen in the middle part of the post-alveolar region. These authors regard *B. fusiformis* as the equivalent of *B. Fleuriausus* of D'Orbigny. But the author of the 'Paléontologie Française' was of a different opinion, and placed *B. fusiformis* of Miller among the many synonyms which he quotes with *B. hastatus* of Blainville. In Pl. XXVI all the different forms of Stonesfield Belemnites as yet discovered are represented ; they all possess a depressed post-alveolar region, and a ventral sulcus reaching far toward the apex, and extending considerably in the alveolar region.

Belemnites Bessinus, *D'Orb.*   Pl. XXVI, fig. 63.

Reference.   *B. Bessinus*, D'Orb., 'Pal. Franç. Ter. Jur.,' p. 110, pl. xiii, figs. 7—13,
   1842.
   B. canaliculatus, Quenst., 'Ceph.,' p. 438, pl. xxix, fig. 7, 1849.

Guard.   Elongate, gently and equally tapering till near the apex, which is subacute;
depressed in all the post-alveolar and part of the alveolar region, nearly circular in the
advanced part of the alveolar region; ventral surface marked by a deep distinct groove
reaching from near the apex to about the last or most advanced of the septa in the
phragmocone.

Transverse section nearly circular at the alveolar apex, much depressed in all the
region behind it.

Dimensions.   Greatest length observed $6\frac{1}{10}$ inches (specimen in Oxford Museum); the
alveolar cavity occupies $1\frac{1}{4}$ inch.   Diameter at alveolar apex $0\cdot6$ inch.

Proportions.   Taking the diameter $v\,d$ at the alveolar apex $= 100$, that from side to
side $= 105$; the axis is excentric and $= 500$; near the apex the diameter from side to
side is to that from back to front as $135 : 100$.

Phragmocone.   In several specimens this part of the Belemnite shell is seen; in one
of my examples it is very well seen.   The angle is from $18°$ to $29°$; the sides are straight;
the back and front very slightly curved; the apex is a spherule; the septa are at the
ordinary distance—about one sixth of the diameter; the axis of the chambered part
visible is half the length of the axis of the guard; the diameter of the largest septum
(crushed) $\frac{1}{2}$ inch.   But single septa occur in the Stonesfield strata which measure 1 inch
across, and are very nearly circular.

Locality.   Stonesfield, Oxfordshire, in the lowest fissile beds of the Great Oolite.

Observations.   The specimens which are represented in Pl. XXVI under the name of
B. Bessinus, agree very fairly with D'Orbigny's figure already referred to, except
that no trace of contraction in breadth appears about the alveolar apex, as in his example,
giving a slightly subhastate figure to the guard.   But a specimen in the Oxford Collection, of the same size as the largest of our figures, exhibits this peculiar outline in a very
slight degree; it also shows, but not very clearly, the double shallow stria which is
mentioned at the retral extremity of the deep, well-defined ventral groove (D'Orb., pl. xiii,
fig. 7).   If to these points of agreement we add the conformity of the angle of the
phragmocone ($20°$ in each), there will be little reason to doubt the agreement of the species.
D'Orbigny obtained his specimen from the Inferior Oolite of Port-en-Bessin (Calvados).

BELEMNITES ARIPISTILLUM, *Llwyd.* Pl. XXVI, fig. 64.

Reference. *Belemnites aripistillum referens,* Llwyd, 'Lithophylacium Britannicum,' No. 1705, pl. xxv, fig. 1705, 1699.

Fusiform Belemnite, Platt, 'Phil. Trans.,' pl. iv, fig. 2, 1764.

B. fusiformis, Parkinson, 'Org. Rem.,' p. 127, vol. iii, pl. viii, fig. 13, 1813.

B. fusiformis, Miller, 'Geol. Trans.,' p. 61, vol. ii, pl. viii, fig. 22; pl. ix, figs. 5, 7, 1826.

B. fusiformis, Morris and Lycett, 'Great Oolite Mollusca,' Part. I, p. 8, pl. i, figs. 6 and 8, 1851.

GUARD. Elongate, fusiform, anteriorly circular, posteriorly depressed, with a deep well-defined furrow on the ventral surface, extending over the alveolus, and reaching to the apex, or near to it.

The young and old agree in general form.

Transverse sections show the outline to be reniform in all the post-alveolar region, the longest diameter being from side to side; in the most advanced part of the alveolar region the section is nearly circular; the axis is excentric and straight.

The longest specimen of the guard, including its expansion over the alveolus, measures $3\frac{1}{4}$ inches; the alveolar portion being $\frac{3}{4}$ inch. Diameter from side to side at the alveolar apex ·21, at the expanded part ·36.

Proportions. Taking the diameter from back to front at the alveolar apex at 100, that from side to side is 120, the axis 1000.

PHRAGMOCONE. Partially observed in several specimens, in none completely.

Locality. Stonesfield, Oxfordshire (*Phillips*) and Eyeford, Gloucestershire (*Buckman*); in the lowest fissile beds of the Great Oolite.

Observations. Morris and Lycett ('Pal. Soc. Monog.') and Morris ('Catalogue of British Fossils') give for a synonym *B. Fleuriausus* of D'Orbigny, 'Terrains Jurassiques Céphalop.,' pl. xiii, figs. 14—18. D'Orbigny himself treats *B. fusiformis* as identical with *B. hastatus*—remarking of this species the unusual distance between the septa, the apex submucronate and free from groove for one third of the apicial length, and the small angle of the phragmocone (11° to 18°). "It is," he says, "without contradiction, the most characteristic species of the lower Oxford Clay, where it constitutes a positive and certain zone." If we take the drawings and descriptions of D'Orbigny for guide, and compare with them the fossils of British localities, we shall find in the Oxford Clay of Cowley near Oxford, close analogies to *B. hastatus,* and in the fossils of Stonesfield equal resemblance to *B. Fleuriausus,* which is quoted from " Great Oolite" at Luçon (Vendée).

16

BELEMNITES PARALLELUS, n. s.   Pl. XXVII, figs. 65, 66.

Reference.—*Belemnites canaliculatus* Quenstedt., 'Cephalop.,' pl. xix, fig. 4, from
beds below *Ammonites macrocephalus;* and *B. fusiformis* of the same
author pl. xxix, fig. 40, from the Great Ooolite of Lahr, in the Rhein-
thal, may probably be of his species.   1849.

GUARD.   Elongate, depressed, except in the advanced alveolar region; fusiform when
young, then becoming hastate or subhastate (old specimens unknown at present).   Ventral
surface marked by a distinct groove, extended forward over the alveolar cavity and backward toward the very acute apex, but terminating in that direction so as to leave free
from groove a length equal to one third of the axis of the guard.

Transverse sections circular across the forward part of the alveolar cavity, depressed
and reniform in all the post-alveolar part, except toward the apex, where they pass from
elliptical to circular.

Dimensions.   From $\frac{1}{2}$ inch to $3\frac{4}{10}$ inches, of which the axis is about 2 inches and
$\frac{8}{10}$ths.

Proportions.   In the oldest example yet observed, the axis is seven times as long as
the greatest post-alveolar breadth, ten times as long as the breadth at the alveolar apex,
and between eleven and twelve times as long as the ventro-dorsal diameter there.
Axis, therefore, 1150; the ventro-dorsal diameter being taken at 100.

PHRAGMOCONE.   In one specimen 24 septa are counted in a quarter of an inch from
the apex, which is terminated by rather a large spherule.   The angle of the phragmocone
is about 28°.

In young fusiform specimens, resembling an oat-grain, the alveolar part is rarely
traceable, the pearly laminæ of the guard having perished, or fracture having
occurred.   In older specimens these white laminæ are covered over by darker and more
solid layers.

Locality.   In Clay coloured on the Ordnance Map as the "Fuller's Earth," between
Great and Inferior Oolite, at Whistle Bridge, near Yeovil (*Ibbotson, Buckman, Wood*);
and Misterton, near Sherborne (*Buckman*).   In the Museum at Strasburg a specimen
marked from the "Fuller's Earth at Oerschingen," above 5 inches long, much resembles
our specimens from Dorsetshire.   In the Oxford Collection are specimens from the Oxford
Clay at Long Marston, near Oxford, which cannot be distinguished from those of Dorset.

EXPLANATION OF PLATE XXI.

Fig.

53. BELEMNITES ELLIPTICUS.

l'.   Lateral view of specimen in the Bristol Institution, from Dundry.

v'.   Ventral aspect of the same, showing the greatly compressed shape.

l''.   Lateral view, showing rugose extremity, and trace of lateral groove.

l'''.   Lateral view, showing worn surface, and groove near the apex.

v'''.   Outline of the ventral aspect.

s'''.   Cross section of the same.

s'.   View of a septum with the siphuncle, showing its unusually elliptical outline.

ϕ'.   Side view of two septa.

ϕ''.   Side view of phragmocone, with the conothecal laminæ.

σ.   Striation on the conotheca; s'', one of the septa of ϕ''; s''', another septum.

PL. XXI.

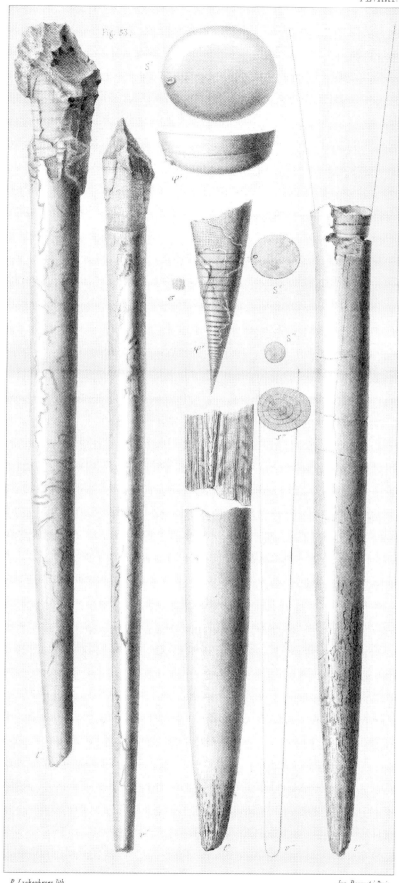

The material originally positioned here is too large for reproduction in this
reissue. A PDF can be downloaded from the web address given on page iv
of this book, by clicking on 'Resources Available'.

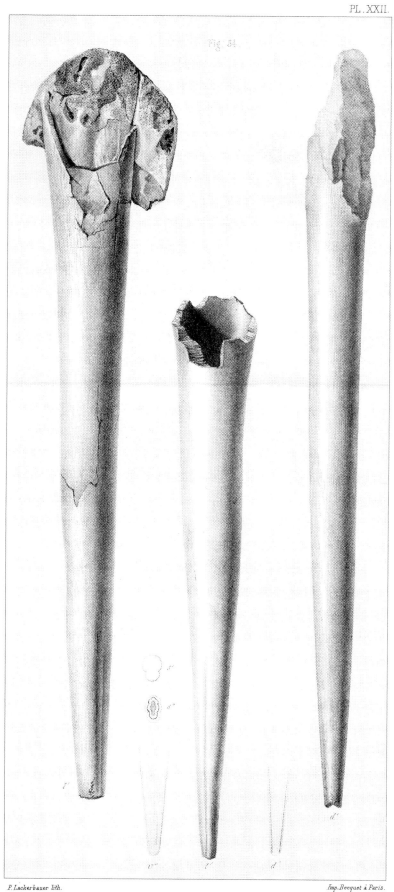

Fig. 34.

P.Lackerbauer lith.

Imp.Becquet à Paris.

EXPLANATION OF PLATE XXIII.

Fig.

55 BELEMNITES GIGANTEUS.

l′.   Lateral aspect.   (From the Inferior Oolite, near Sherborne, Dorsetshire.) In the Collection of Mr. Reed, Salisbury.

The specimen was cracked and displaced after being enclosed in the rock.

56 BELEMNITES QUINQUESULCATUS.

l″.   Lateral aspect, with six-channelled apex (From near Sherborne.) In the Collection of Mr. Reed.

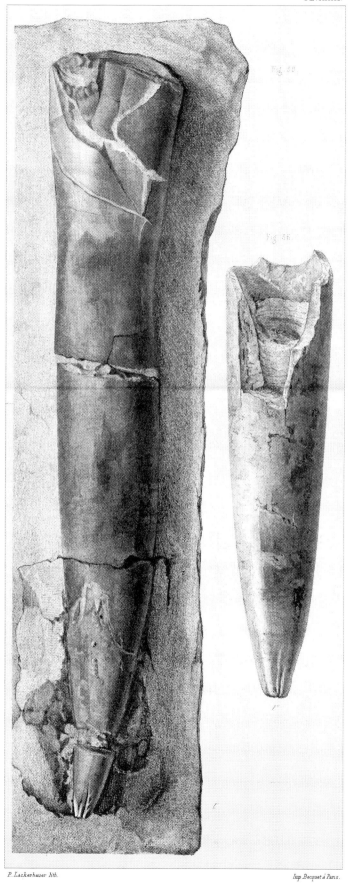

The material originally positioned here is too large for reproduction in this reissue. A PDF can be downloaded from the web address given on page iv of this book, by clicking on 'Resources Available'.

EXPLANATION OF PLATE XXIV.

F_{IG}.

57 B<small>ELEMNITES</small> <small>QUINQUESULCATUS</small>.

l'.   Lateral aspect. (From the Grey Limestone of White Nab, Scarborough.)
By erosion at the point the grooves appear.

v'.   Ventral aspect, striato-sulcate.

d'.   Dorsal aspect, showing the two lateral and one short dorsal groove.

l''.   Lateral aspect, with grooved apex, and the phragmocone *in situ*.

d''.   A specimen, seen dorsally, to show the lateral grooves, and on the phragmocone the hyperbolic dorsal arcs.

ϕ''.   Striation on the conotheca.

l'''.   Fragment seen laterally, showing by accidental fracture the interior
apicial grooves.

s.   Cross section in the alveolar region.

s', s'', s''', s^{iv}.   Cross sections of the sheath. In s', the interior laminæ show
lateral grooves : in s'', the terminal grooves are marked by seven tinted
radial parts : in s^{iv}, several grooves appear.

Fig.57.

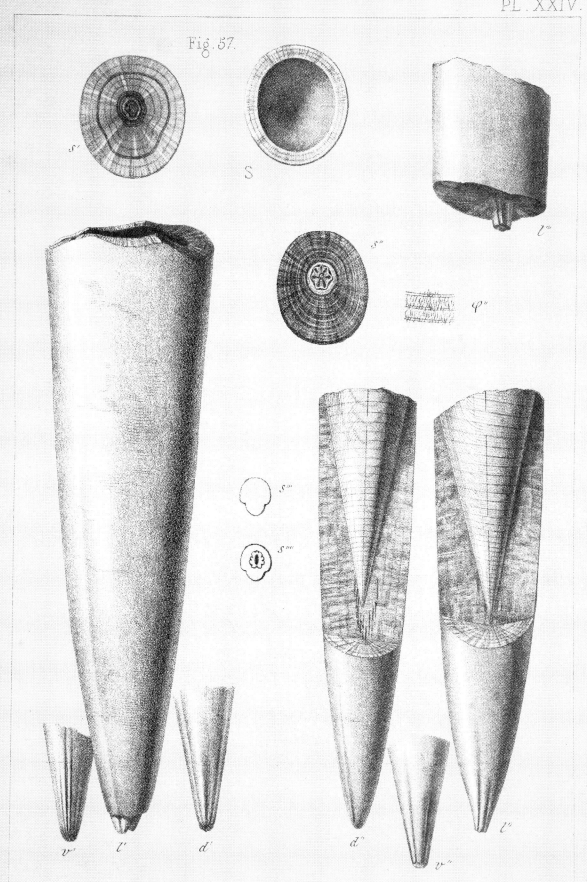

P. Lackerbauer lith.

Imp.Becquet à Paris.

EXPLANATION OF PLATE XXV.

Fig.

58. BELEMNITES APICICONUS. (From Yeovil.)

    v.   Ventral aspect, showing an unusually definite posterior ending to the groove, and its continuity over the alveolar cavity. Specimen in the Oxford Museum, from Inferior Oolite, Yeovil.

    l.   Lateral view of the same : s, section across the alveolar cavity ; *v'*, frustum seen vertically to show the distinctness of the groove.

59. BELEMNITES BLAINVILLII. Long variety. (From Sherborne.)

    v'.   Ventral aspect, showing the groove reaching nearly to the apex, and dying out on the alveolar cavity.

    v".   Another example, showing the groove certainly extending over a part of the alveolar space.

    l.   A side view.

    s.   Section across the alveolar cavity; *s,* section across the guard at the alveolar apex.

60. BELEMNITES BLAINVILLII. Shorter variety. (From Sherborne.)

    v'.   Ventral aspect, the groove reaching to the phragmocone.

    s'.   Cross section.

    v"   Another specimen, showing the groove extended further forward.

    s".   Cross section of the same ; *l",* lateral view.

    v".   Ventral aspect of a specimen, where the groove does not quite reach the alveolar cavity ; *s"',* cross section of the alveolar cavity.

61. BELEMNITES CANALICULATUS.

    v'   Ventral aspect   of a small specimen from Dundry.
    l'.   Lateral aspect

    v".   Ventral aspect of a larger specimen, a little bent, from Dundry.

    s".   Cross section of the sheath ; *s",* section across the alveolar cavity.

62. BELEMNITES TERMINALIS, n. s.

    l.   Lateral view.

    v.   Ventral aspect.

    s.   Section across the alveolar chamber.

    s.   Section across the guard.

PL. XXV.

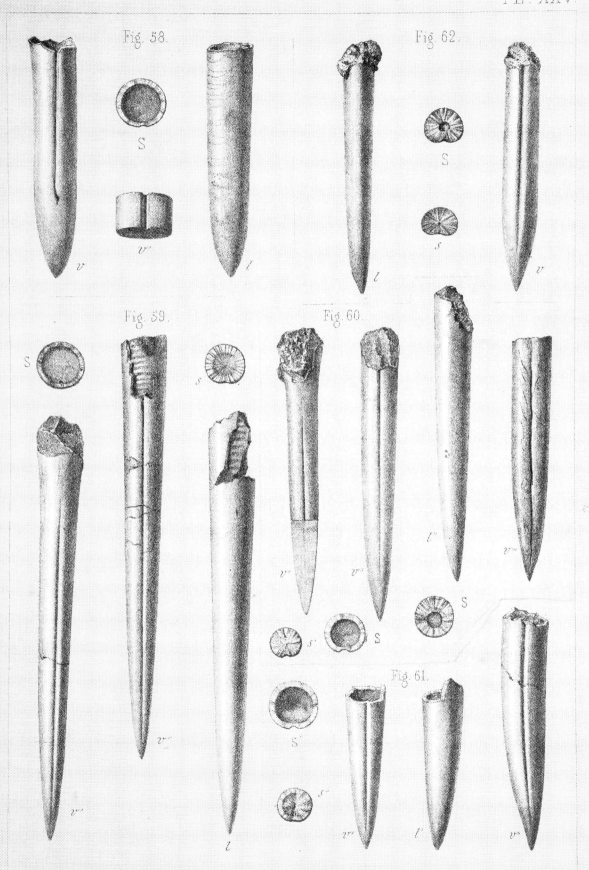

Fig. 58.

Fig. 62.

Fig. 59.

Fig. 60.

Fig. 61.

P. Lackerbauer lith.

Imp. Becquet à Paris.

EXPLANATION OF PLATE XXVI.

Fig.

63. Belemnites Bessinus. (From Stonesfield.)

v'.   Ventral aspect of a specimen, showing the phragmocone *in situ*, and the groove continued over it.

s', s''. Cross sections of the sheath.

v''.   Ventral aspect of a large specimen in the Oxford Museum.

v'''.   Similar view of a small specimen.

v^{iv}.   Another specimen of intermediate size.

l^{iv}.   Lateral view of the same.

s^{iv}.   Cross section of the alveolar chamber.

64. Belemnites ari-pistillum. (From Stonesfield.)

v'.   Ventral aspect with phragmocone, *in situ*.

s'.   Sections across the guard.

v''.   Small specimen ; s'', its cross section.

v'''.   Specimen of one somewhat larger.

v^{iv} Largest example which has been observed by the author. In Mr. J. Parker's Collection.

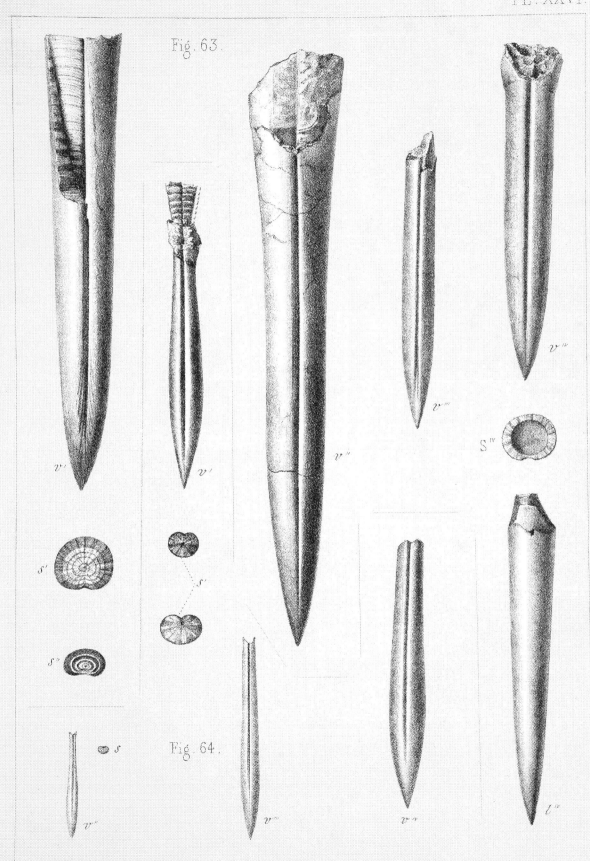

Fig. 63.

Fig. 64.

P. Lackerbauer lith.

Imp. Becquet à Paris.

EXPLANATION OF PLATE XXVII.

Fig.

65. Belemnites parallelus, n. s.   (Upper series on the Plate.)

v'.   Ventral aspect of a fragment, showing the alveolar cavity, and an abnormal
expansion of the groove: from Misterton, near Sherborne.

l'.   Lateral aspect of a fragment from Misterton.

l''.   Lateral view of the largest individual from Misterton.

v''.   Ventral aspect of the same.

s''.   Sections across the alveolar cavity.

s''.   Sections across the sheath.

66. Lower series on the Plate.

l^{iv}.   Lateral views of specimens from Whistle Bridge, near Yeovil.

d^{iv}.   Dorsal view, and v^{iv} ventral view of the same.

s^{iv}.   Sections of the guard of the same.

l^{v}, d^{v}, v^{v}.   Young specimens.

v'''.   Shows the relation of the young to the larger figures.

ϕ''.   Restoration of the phragmocone.

ϕ'''.   Longitudinal section (by splitting), showing the phragmocone.

s'''.   Cross sections of the alveolar cavity.

Fig. 65.

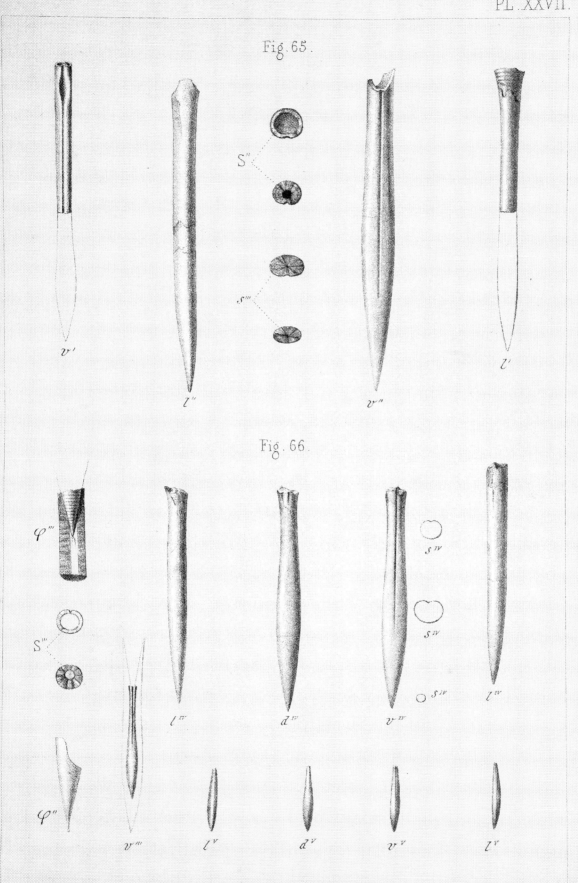

P.Lackerbauer lith.

Imp.Becquet à Paris.

A MONOGRAPH

OF

BRITISH BELEMNITIDÆ.

BY

JOHN PHILLIPS,

M.A. OXON., LL.D. DUBLIN, LL.D. CAMBRIDGE, D.C.L. OXON., F.R.S., F.G.S., ETC.,
PROFESSOR OF GEOLOGY IN THE UNIVERSITY OF OXFORD.

PART V,

CONTAINING

LONDON:

PRINTED FOR THE PALÆONTOGRAPHICAL SOCIETY.

1870.

PRINTED BY

J. E. ADLARD, BARTHOLOMEW CLOSE.

BELEMNITES OF THE OXFORD CLAY.

In passing upward from the thin-bedded rocks of Stonesfield, we find few or no Belemnites for a considerable thickness of the Oolites. Through the whole series of the Great Oolite, Bradford Clay, Forest Marble, and Cornbrash, Belemnites, if ever found, are very rarely seen in the South of England. In the North of England, the doubtful "Grey Limestone," as I termed it, of the Yorkshire Coast, contains Belemnites, but they are of the type of Inferior Oolite, and with *Ammonites Blagdeni, Am. Humphreysianus,* and *Am. Parkinsoni,* must be held to carry that rock to the "Lower Badonian" stage. Is the Great Oolite of the South of England wholly devoid of Belemnites, except in its lowest member, the Stonesfield slate? I can only reply that no specimen has occurred to my personal observation. Does the Bradford Clay contain any Belemnites? Only one notice is on record, and that is in the now rarely seen volume, published by W. Smith, under the title of 'Stratigraphical System of Organized Fossils,' 1817. In that work, page 79, occur these words:—"Multilocular bivalves. Belemnites small, slender; Stoford." As my boyish hand wrote the words—the place being familiar to me, I have no reason to doubt the accuracy of the record. The specimen was transferred to the British Museum. No Belemnite is mentioned in the Forest Marble beds, nor, so far as I now remember, has any one been quoted from the Cornbrash, except by error. In the first edition of my work on the 'Geology of the Yorkshire Coast,' 1829, I remarked (p. 145) "No Belemnites have been found in the Cornbrash of Yorkshire;" and again (p. 146), "The Cornbrash is the only conchiferous stratum in the eastern parts of Yorkshire from which Belemnites are excluded." In consequence of some notice reaching me of a specimen found in the Cornbrash of Yorkshire between 1829 and 1835, I modified the expression in the Second Edition, so as to call attention to the extreme rarity of the occurrence. If any Palæontologist whom these remarks may reach should find himself able to furnish me with specimens of Belemnites from beds between the Cornbrash and Stonesfield slate, of any part of England, he would oblige me much by a sight of them.

There being then, as appears, this great blank in regard to Belemnites (the remark is almost equally good for Ammonites, but in this case we must exclude the Cornbrash), through a considerable range of conchiferous strata, it becomes a matter of great interest to compare the several Oxonian forms which now appear, with the numerous Badonian species which have disappeared. Are these the same species matured in some other part of the sea, modified there through a long succession of transmitted forms, and again brought into the Oolitiferous ocean? We may consider this question after the facts have been collected and studied.

Among the Belemnites of the Oxford Clay and the Kelloway Rock (a sandy member which is not seldom absent from the section), four principal forms appear to have reached maturity in the area of England, which may, for convenience, be termed ' *Hastati,*' including *B. hastatus* of Blainville; ' *Canaliculati,*' including *B. sulcatus* of Miller; ' *Tornatiles,*' including *B. Puzosianus* of D'Orbigny; and ' *Excentrici,*' including *B. abbreviatus* of Miller. When we endeavour to trace the history of these several forms, from the youngest examples, we experience in more than the usual degree the difficulty of obtaining a series of all ages for any one of the species.

However carefully we may collect, in many favourable localities, it is nearly impracticable to fill up all the terms of the series; and though scores of young ones have been collected by my own hands from different localities, it is only in a few instances, and by the aid of my pupils, that I have succeeded in proving to my own satisfaction the real progress of these several forms towards maturity. Nor does the method of examination by sections of the older individual succeed in this case so well as in some others, because of the frequently very close texture of the sparry substance, and its more complete condensation into an amber-coloured mass, than is usual in the earlier deposits. For this reason the form of the very young shell cannot always be even approximately known by examining polished or natural sections. As far as can be judged from these sections, however, there is reason to think that most of the Belemnites of the Oxford Clay began life in a more or less hastate, or else lanceolate, shape; and this seems to be confirmed by the fact, as I believe it to be, that no very small specimen has ever been observed in the Oxford Clay or Kelloway Rock in England, except in one of these shapes. Extending our view to Scotland, we find a somewhat different result. The Belemnites of the Cromartie coast have been collected very successfully at Eathie and Shandwick by Lieut. Patterson, of Ripon, to whom I am obliged for the sight of his fine series, and for photographs of many specimens.

Two species at least have there arrived at maturity; one, a peculiar elongated spicular Belemnite, whose guard sometimes reaches the length of ten inches, is found in Mr. Patterson's series of all sizes down to one inch: it is only in this very small and very slender specimen that any approach to a fusiform shape of the guard can be recognised, and then only in a very slight degree. Another of these species makes an approach to *B. sulcatus* of Miller, and is longitudinally grooved up to the point, at least in all the smaller specimens (Shandwick).[1] There is nothing in the smallest of these at all comparable to the clavate forms common in the Oxford Clay of England, though a slightly hastate shape can be recognised among them. The strata from which these Belemnites come have been called Lias, but what *Ammonites* and *Conchifera* I have seen from them are of the Oxonian type of life.

Two of the four Oxonian groups have been already mentioned in the Badonian

[1] In Lieut. Patterson's Collection is one specimen of a decidedly Liassic Belemnite of the group of *B. elongatus* (Miller), which is placed among the Shandwick fossils.

Oolites (*Hastati* and *Canaliculati*); the others (*Tornatiles* and *Excentrici*) now first make their appearance, to replace (perhaps we may say) the *Gigantei*, which they, however, resemble in no particular except size.

BELEMNITES HASTATUS, *Blainville.* Pl. XXVIII, figs. 67—70.

> *Reference.* Var. *a.* *Belemnites hastatus*, Blainville, 'Mém. sur les Bélemn.,' p. 71, pl. 2, fig. 4, 4 *a*. 1827.
> *Belemnites semihastatus*, Blainv., 'Mém. sur les Bélemn.,' p. 72, pl. ii, fig. 5, 5*a*—5*g*. 1827.
> *Belemnites gracilis*, Phillips, 'Geol. of Yorkshire,' vol. i, p. 138, pl. v, fig. 15. 1829.
> *Belemnites hastatus*, D'Orb., 'Terr. Jurass.,' p. 121, pl. xviii, fig. 1, 9. 1842. (Exclude some of the synonyms.)
> *Belemnites hastatus*, Quenstedt, 'Cephalopoda,' p. 442, pl. xxix, fig. 27—39. 1849.
> *Belemnites semihastatus rotundus*, Quenstedt, 'Cephal.,' p. 440, pl. xxix, fig. 8—11. 1849.
> *Belemnites semihastatus depressus*, Quenstedt, 'Cephal.,' p. 440, pl. xxix, fig. 12—18. 1849.

GUARD. Very elongate, smooth, hastate, with an acute apex (by decay of laminæ about the alveolar apex it becomes fusiform); in all stages of life depressed and expanded laterally in the post-alveolar region; cylindrical or somewhat compressed about the alveolus; ventral surface marked by a distinct groove, which is extended forward over the alveolar cavity, and backwards toward the apex, about half the length of the axis of the guard, so as to leave much of the expanded part free from groove, or with merely a faint indication or trace of groove. In young specimens this part appears swollen on the ventral aspect (see fig. 69, 2^2, 2^4).

Transverse sections show the laminæ of the guard thickening on the dorsal aspect of the alveolar cavity, so as to make the section circular, or a little oblong there; the sectional outline is depressed and reniform in the post-alveolar region, till the bulbous part is reached, which has an oval section gradually growing circular toward the conical or slightly sub-mucronate point. The axis is nearly central. In some excellent specimens are small faint longitudinal bent or sigmoidal furrows one on each side of the guard. They begin nearer to the ventral than the dorsal surface, and bend upward before losing their distinctness (figs. 68*l'*, 67*l*).

Dimensions. In the Oxford district within a few miles of the city, the clay-pits have been well searched for these Belemnites, and with great success as far as the young forms are concerned. Only one specimen has yet been met with corresponding in size to the figures of D'Orbigny and Quenstedt already referred to. This is in the collection of Mr. James Parker, from the Cowley clay pit at its deepest part, about 80 feet below the Calcareous Grit. This specimen has a total length of 6·75 inches; greatest breadth 0·65; least 0·40; greatest depth 0·60; least 0·40; axis = 5·5 inches. Between this fine and solitary specimen and very many examples 2¼ inches long, no intermediate magnitudes have yet, near Oxford, been found; the smallest specimens, like oat grains, are about half an inch long, and then by decay of the laminæ about the alveolar apex acquire the aspect of the so-called 'Actinocamax' of Miller.

Proportions. In young specimens the axis is about seven times as long as the greatest post-alveolar breadth, and about ten times as long as the breadth at the alveolar apex : in an old specimen, the axis is fourteen times as long as the alveolar breadth, and nine times as long as the greatest post-alveolar breadth.

It was with great pleasure that I received from the Oxford Clay of Eyebury, near Northamptonshire, a specimen found by Mr. Leeds (Pl. XXVIII, fig. 68), which happily fills up the blank in the history of the species, by a form of intermediate magnitude, not elsewhere recognised. We are thus assured of the persistence of the hastate form in this species through all stages of growth yet observed, from the very young to the apparently full-grown individuals.

PHRAGMOCONE. Very few indications of this part of the fossil have been as yet seen by me. D'Orbigny, who had fine specimens at his disposal, figures the phragmocone of one ('Terr. Jurassiques,' pl. viii, fig. 1) in the sheath, and represents it as having an angle of about 15° (in the description it is said to be 11° to 18°), with chambers whose diameter is only four times their depth. Quenstedt ('Cephal.,' pl. 29, fig. 8 *a*, 9) presents the phragmocone of *B. semihastatus rotundus* (regarded as a variety of *B. hastatus* by D'Orbigny), with septal intervals equal to one fifth of the diameter, and an angle of 13°. These may be regarded as good characters for discriminating between this specific group and that of *B. ari-pistillum* of Stonesfield. The septa are more nearly round than in the figure of D'Orbigny.

Localities. In Oxford Clay, Weymouth; in the middle part of the clay north of the town; on the shore in the upper part of the clay; and south of the town. Oxford, in the lower or middle part of the clay, with *Ammonites Duncani*, at Summertown, one mile to the north; in the upper part of the clay at Cowley Field, half a mile to the south-east; and at Long Marston, in the upper part of the clay, one mile to the north-east (*Phillips*). Eyebury, near Peterborough, in the lower part of the clay (*Leeds*). St. Ives (*Walker*). Scarborough Castle Hill (*Phillips*). In Calcareous Grit, Scarborough (*Bean*).

Observations. D'Orbigny collects under one title the two fossils to which Blainville assigned the names of *Belemnites hastatus* and *B. semihastatus*. The differences between them were far from clear in the earlier author's descriptions or figures. Quenstedt retains the distinct names, and is, in this respect, in agreement with other German writers. Collections in Germany follow this model in their arrangements. D'Orbigny joins to these *Belemnites fusiformis* of several writers, *B. ferrugineus* of Voltz, and *B. gracilis* of Raspail.

In considering the varieties to which the species seems liable, we find among English specimens, of lengths from half an inch to three inches, some differences in the general shape of the guard, which in some specimens is elegantly hastate (fig. 69, v^3), in others more expanded and recurved at the apex (fig. 69, l^2, v^2), in a few bulbiform (fig. 69, v^6, v^7), in one deformed (fig. 70, μ). The ventral groove is generally absent from half of the length of the axis of the guard; in rare cases (fig. 69, v^7) it is interrupted, sometimes it leaves more than half of the length of the axis of the guard smooth. I do not observe lateral furrows on any of the small specimens, and it seems rarely absent from any of the larger ones. In the older form, which exceeds six inches in length, the undulated, or somewhat oblique faint double furrow, is traceable nearly as D'Orbigny has represented it ('Terr. Jur.,' pl. xviii, fig. 4).

BELEMNITES HASTATUS, var. BULBOSUS. Pl. XXVIII, fig. 69, v^6, v^7.

The variety to which attention is now called, is more than any other remarkable for the retral expansion of the guard and the swollen ventral outline of the expanded part. In eleven specimens before me, including individuals from ¾ inch to 2¼, the characters were nearly uniform. The broad part of the guard is about ⅓rd of the axis from the apex; the groove usually terminates at half the length of the axis from the apex; but in one specimen (fig. 69, v^7) it is interrupted and farther extended.

The sections are nearly round in and for a small space behind the alveolar portion, but everywhere further back they are elliptical. The laminæ over and a little behind the alveolar space are, as usual, pale and less calcareous than in the more solid part of the guard.

Locality. Specimens of this very interesting form have been forwarded to me by Mr. J. F. Walker, of Sidney Sussex College, Cambridge, from the Oxford Clay of St. Ives, Cambridgeshire, where they are accompanied by *B. Puzosianus* of D'Orbigny, and *B. sulcatus* of Miller.

I have examined in foreign collections a considerable number of specimens called *B. hastatus* and *B. semihastatus*, and considered the figures which are given as representing them. I am unable to perceive differences among them or among the comparatively few English examples of full size, such as to require the employment of more than one specific name. At the same time there are differences; some have a nearly circular section across the expanded part of the guard, others a depressed contour there; similar variations occur in the alveolar region. To the former the 'variety-name' of *rotundus* has been assigned, to the latter *depressus*. The somewhat flexuous lateral groove is absent in some and present in other examples not otherwise differing.

On a group of Belemnites, including *B. canaliculatus*, Schlotheim (in part); *B. sulcatus*, Miller (in part); *B. Altdorfensis*, Blainville *B. absolutus*, Fischer; *B. Beaumontianus*, D'Orbigny.

The canaliculated Belemnites above referred to are frequent in the Oxford Clay, and specially toward the lower part of it, as it occurs in England. They are found in the vicinity of Oxford, associated with *Ammonites Duncani*, in the parallel of the Kelloway Rock, or nearly so, for that rock is hardly traceable in this quarter. In the corresponding clay of Weymouth, Belemnites are found of the same general character, while at St. Neot's specimens occur which cannot in the least particular be distinguished from Oxford specimens.

Miller, while examining the Oxford Collection, certainly referred the channelled Belemnite of the neighbourhood to *B. sulcatus* ; but a frequent application of this name is to a species of the Lower Oolite, such as *B. apiciconus*. We find in Schlotheim *B. canaliculatus* corresponding to *B. sulcatus* of Miller, and, like it, including forms from the Inferior Oolite and the Oxford Clay. Blainville rightly separates them, and assigns to his *B. Altdorfensis* one of Miller's figures (pl. viii, fig. 5, ' Geol. Trans.,' 2nd series, vol. ii), and a part of *B. canaliculatus* of Schlotheim. Quenstedt employs the general title of *B. canaliculatus* for all these forms, and includes in it the Stonesfield fossils referred to *B. Bessinus* by Morris and Lycett.

Belemnites having the same general character occur in the Oolitic series of Russia, with Ammonites of the Oxford Clay; and similar forms have come to my hand from the Himalaya.

Among all these fossils there is so much of resemblance that in the sense of the term species, as it was employed by the earlier naturalists who thought with Linnæus, they might be classified under one title, such as *B. canaliculatus*, the earliest on record, as Quenstedt does. But this title is equally claimed for the grooved Belemnites of the Bath Oolite series, which contain several very distinguishable and characteristic forms.

B. Altdorfensis of Blainville is supposed by this author to be identical with *B. canaliculatus* of Schlotheim and *B. sulcatus* of Miller, and is quoted from the ferruginous Oolite of Curey, near Caen.

D'Orbigny disposes of the perplexity of this nomenclature by instituting a new species, *B. Beaumontianus*, which he refers to the Lower Oxford Clay of Vaches-Noires. A fossil, corresponding to his figure, occurs at Loch Staffin in the Isle of Skye, according to Prof. E. Forbes. No other locality is given by Morris.

Upon the whole I am disposed to preserve the name which Miller certainly imposed on the long-grooved fossils from the Oxford Clay; the more so as it will be seen that hardly any examples fit so exactly with the figure of *B. Beaumontianus* given by D'Orbigny as to render that a good general type of a variable species.

Few Belemnites appear to have had so large a distribution in time and space as the group allied to *B. sulcatus* of Miller and *B. canaliculatus* of Schlotheim. From the base of the Inferior Oolite to the middle of the Oxford Clay they are generally recognised in Europe; specimens much like our examples from Oxford Clay are abundant in the country south of Moscow; others come to us from the Himalaya[1], from Cutch[2], South Africa[3], New Zealand,[4] and Queensland.[5]

BELEMNITES SULCATUS, *Miller.* Pl. XXIX, XXX; figs. 71—75.

Reference. *Belemnites sulcatus*, Miller, ' Geol. Trans.,' 2nd series, vol. ii, p. 59, pl. viii, fig. 5 [excl. fig. 3 and 4]. 1823.
Belemnites Beaumontianus, D'Orb., ' Pal. Franç., Terr. Jur.,' p. 118, pl. xvi, fig. 7, 11 (on the plate it is called *B. Alldorfensis*). 1842.

GUARD. Subcylindrical or conical in the alveolar region, more or less depressed in the post-alveolar region and deeply grooved; the groove interrupted or expanded toward the apex, and gradually ceasing about the alveolar summit. Outline nearly straight on the ventral, more curved on the dorsal aspect; apicial region tapering, surface smooth or granulated. Sections show the axis to be nearest the ventral face, very excentric, and somewhat recurved. Near the apex the sections are almost circular or a little oblong.

Greatest length observed 5·5 inches; and of this the axis occupied 3 inches. Greatest diameter in this specimen 0·85, in a stouter specimen 1 05.

Young. The very young form was more or less hastate (fig. 73 v^1). Somewhat advanced in age is the very rarely seen form fig. 73 *v*, from near Oxford; next we have fig. 72 and 71 *v*, differing from full-grown specimens only in greater slenderness.

Proportions. The normal diameter (not counting the groove) being taken at 100, the transverse diameter of the alveolar apex is, in full grown individuals, 108; the axis under 300; the ventral radius 40; the dorsal 60. In young specimens the axis is 500.

PHRAGMOCONE. Slightly arched, very obliquely inserted; septa nearly circular, unusually approximate, their depth being about one eighth or even only one ninth of the diameter in the anterior part; sphericle distinct and rather large; angle 22°. The concave surface, within the septal edge, is a portion of a sphere, measuring 90° across.

Observations. There is some variety in the sections of the guard; some specimens showing more depression than others; in some the groove is broader, in others it is deeper; in a few the groove expands a little toward the apex (fig. 74 *v*), and also expands on the surface over the alveolus (fig. 75 *v'*); in some there are one or two lines parallel to the edges

[1] Specimen in my possession. [2] Sowerby, ' Geol. Trans.,' 2nd Ser., vol. v, p. 329.
[3] Tate, in ' Geol. Soc. Journal,' 1867, p. 151. [4] Hochstetter, Novara-Expedition.
[5] Specimens in the Collection of Mr. Charles Moore.

of the groove (fig. 75 *v′* and *v″*). There is often a marked increase of depth and definition of the groove for half the length of the guard, measuring back from the alveolar region, as if in that part was a fissure (fig. 71, *v′*, *v*). A faint intimation of the groove can almost always be traced to near the apex.

Specimens occur with an external sheath of white fibrous matter, rough on the outer surface (Pl. XXIX, fig. 72 *m*). One might fancy this to be a periostracon or capsule, but it is, I believe, really a concretionary deposit. The shell is sometimes granulated (Pl. XXIX, fig. 72 *m*).

In figure 10, pl. xvi, of the ' Pal. Franç.' the outline of the alveolar cavity, erroneously represented as somewhat transverse, should have been very nearly circular. The Oxford specimens are never so much depressed in the post-alveolar region as in fig. 9 of the same plate.

The axis of the guard of this Belemnite, in some specimens obtained from the Oxford district, is hollow for a part of the length, as if the apices of the young laminæ of the guard were, during life, removed, so that a sort of pipe, partially interrupted at intervals by the edges of these laminæ, extended inwards from the perforated apex. Afterwards the sheaths successively formed covered them completely, and were not perforated. In some specimens (fig. 71 s′, ʌ′, ʌ″) the axial canal is very narrow for a certain space above the alveolar cavity, then it enlarges in a fusiform shape, and again contracts to the mere line of junction of the opposite guard-fibres. This curious appearance will be further considered in connection with *B. abbreviatus*.

Another very curious fact is observed in several of these fossils. On the ventral aspect, internally, are one or two cavities extended lengthways, through the substance of the guard, from a little in front of the alveolar apex to a greater distance behind. An explanation is found by the aid of cross sections: for these, taken a little behind the alveolar apex, show the cavities in question to be formed by the peculiar inflexion of the laminæ of the guard on the ventral aspect. This inflexion becomes remarkable only after a certain age; thenceforward grows continually deeper and deeper, always producing a groove, and sometimes by the contraction of this groove completely or partially enclosing longitudinal canals.

Fig. 71 *s* shows the arrangement of the laminæ round the axis of the guard in conformity with this description. The axis is not tubular in this instance. The laminæ of the guard are crossed by the fibres nearly at right angles to the surface, and as this is a curve of contrary flexure about the ventral aspect, the fibres assume there remarkably arched directions. In these sections glistening dagger-shaped parts are present—they are merely the obliquely truncated prismatic cells of the so-called fibrous structure. It may be well to mention, that the specific gravity of most Belemnites (2·8) agrees with that carbonate of lime called arragonite, and not with ordinary calcite.

The student of Homology will not fail to remark the analogy which this repetition of deep folds on the ventral aspect of *Belemnites sulcatus* offers to the more regular groove on the same aspect in *Belemnitella*. The groove of the latter group, however, is only on the alveolar region, and reaches to its anterior edge, which is emarginate in consequence, while that of *B. sulcatus* belongs to the posterior part, dies out on the same region, and ceases nearly opposite the alveolar apex. The canaliculated axis occurs in some examples of *B. Bessinus* of Stonesfield and in *B. lateralis* of Speeton, but I have not yet seen it in any Liassic species.

Belemnites perforatus, Voltz (Pl. VIII, fig. 2), from Cretaceous beds at Osterfeld, is canaliculated for the whole length of the axis of the guard; and specimens of *B. quadratus* and *B. mucronatus* from the Chalk frequently show this peculiarity, or else a condition of the central parts which suggests their easily acquiring it.

Locality and distribution. Weymouth has yielded characteristic specimens of this species from the Oxford Clay, but they seem not to be plentiful there. I found only two or three fragments in the clay on the shore north of Weymouth, mixed with hundreds of the young forms of a hastate Belemnite. It is not mentioned among the fossils of the Oxford series known to Smith, who figures and describes the longer Belemnite known at Chippenham as *B. Owenii*. About Oxford we find it rather frequently, especially towards the middle and lower part of the clay deposit, with *Ammonites Duncani*, while in the upper part *B. Owenii* and *B. excentricus* occur more frequently, with *Ammonites vertebralis*. The young forms are very rare in these parts. Near St. Neot's, again, they occur with *Ammonites Duncani*, but not plentifully, as I find by Mr. Walker's communications, and again near Peterborough, as I learn from Mr. Leeds. I doubt the occurrence of the species in Yorkshire, and regard the mention of it in the first edition of my work on the geology of that county (1829) as requiring confirmation.

The locality of D'Orbigny's fossil is thus noticed:—"Elle a été recueillée par M. Tesson dans les marnes Oxfordiennes des Vaches-Noires (Calvados); elle parait y être rare."

ON A GROUP OF BELEMNITES ALLIED TO *BELEMNITES PUZOSIANUS* OF D'ORBIGNY.

In 1816 William Smith figured, in his 'Strata Identified,' on the plate of Oxford [Clunch] Clay fossils, a long subcylindrical Belemnite from Dudgrove Farm, in Wilts. In his 'Stratigraphical System' (1817) the fossil is described as "large, squarish, quickly tapering to the apex; diameter one inch at the largest end, length four or five inches." The figure referred to represents the guard almost complete, with the alveolar cavity exposed. I remember the specimen, which is now in the British Museum. Some years later the species was recognised by my great relative in a fossil of the Kelloway Rock in Yorkshire, to which I gave the name of *B. tornatilis*. Of this I had seen specimens when the first edition of my work on the 'Geology of Yorkshire' was published (1829), and described the fossil as elongated. In the second edition it was named, with an equally brief description, but no figure (1835). In 1844 the rich deposit of

18

Chippenham had yielded its treasures, and Belemnites of the same general aspect, with considerable portions of the phragmocone, and even extensions of the conotheca, had furnished to Prof. Owen the materials for a valuable essay on the structure of the shell and the relations of the animal ('Phil. Trans.,' 1844). The fossil was named by Mr. Pratt *B. Owenii.*

The same great record of science contains, in the volume for 1848, another "Essay on the Belemnites of Chippenham," by Dr. Mantell, in which the figures represent a variety of important facts previously unobserved. The Belemnite which he examined is here called *B. attenuatus*—a name long before appropriated to a species found in the Gault, which had, however, been referred to a new genus, *Belemnitella.* D'Orbigny makes known to us a very similar form of Belemnite, also from the Oxford Clay, to which he gives the name of *B. Puzosianus.* Finally, Mr. Morris, in his excellent 'Catalogue of British Fossils,' 1854, employs the term *B. Owenii,* giving *B. Puzosianus* as a synonym, and under *B. tornatilis* proposes the question if it be not identical with *B. Owenii.*

The natural group thus noticed consists of Belemnitic guards of more than the usual length, with a generally cylindrical aspect, more or less compressed; always marked by a depression, often by a conspicuous groove, from the apex along the ventral surface for a third or half the length of the axis.

To the forms best known must be added one or two more from the midland district of England, and as many from the Oolitic series of the coast of Cromarty. These are inordinately long, but in other respects correspond in general character with the more usual species.

Looking back upon earlier groups of Belemnites, we find nothing so much like these as the long, somewhat compressed forms allied to *B. tripartitus* (see Pl. XI, fig. 28), in the Upper Lias. But all those Liassic forms have lateral grooves near the apex, often very conspicuous; these of later ages, in the Oxonian strata, never.

In regard to the synonymy, there can be little doubt about preserving Mr. Pratt's name, *B. Owenii,* for the whole group; *B. Puzosianus,* D'Orbigny, having certainly to be associated with it, as a variety.

BELEMNITES OWENII, *Pratt.*   Pls. XXXI, XXXII, figs. 76—81.

Reference.   *Belemnites* (unnamed), Smith, 'Strata Identified,' 1816, and 'Strati-
                    graphical System,' p. 55, 1817.
                 B. tornatilis, Phillips, 'Geol. of Yorkshire,' vol. i, ed. 2, 1835 (no
                    figure).
                 B. Owenii, Pratt, 'Phil Trans.,' 1844.
                 B. attenuatus, Mantell, 'Phil. Trans.,' 1848.

B. Owenii, Quenstedt, ' Cephalop.,' pl. xxxvi, f. 9, 1849.

B. Puzosianus, D'Orbigny, ' Pal. Franç., Terr. Jur.,' p. 117, pl. xvi, f. 1—6, 1860.

This frequent species, or group of species, varies much in several important characters. The degree of compression is by no means uniform, but I have seen no example of *alveolar* compression approaching to that represented in D'Orbigny, fig. 4, pl. xvi. All my specimens have in that part a slightly elliptical section. In some the sides of the guard are flattened or a little grooved (Pl. XXXI, fig. 77); the apical region is sometimes unusually compressed, but generally follows the sweep of the sides ; the ventral groove near the apex varies from little more than a mere flattening (Pl. XXXII, fig, 79) to a broad furrow (fig. 76, *v'*), a sharp short rift (fig. 77, *v'*), a narrow groove (fig. 76, *v''*, and fig. 78), and a deep lengthened canal (figs. 80 and 81). The general figure, always long, varies in the proportion of length to diameter.

GUARD. Very long, subcylindrical, more or less compressed, tapering evenly to a point, grooved on the ventral aspect from the apex through one third or more of the length of the axis ; in perfect specimens this groove is often bistriated, or somewhat sharply bordered.

Sections show the axis placed nearer to the ventral surface ; in young specimens the compression is considerable, growing less with age ; there is sometimes a distinct lateral flattening on the middle part of the guard.

Greatest length observed, in specimens from St. Neot's, 10 inches, of which the axis is 6 inches ; the diameter at the alveolar apex 1 inch. In another the diameter at the alveolar apex is $1\frac{1}{4}$ inch.

In a very young state the pearly laminæ about the alveolar apex are sometimes decomposed, and the guard assumes the delusive shape of ' Actinocamax.' A drawing has been shown to me in which this fusiform guard, or ' ossicle,' is represented as separated from the ' nucleus' of the phragmocone, but I have seen no specimen of the kind. Prof. Owen figures (' Phil. Trans.,' 1844, pl. ii, fig. 4) a very young individual, with the guard and alveolar chamber in their ordinary relations. The guard is in this state shorter in proportion than in after-life.

Proportions. In a full-grown specimen from St. Neot's the diameter from back to front, at the alveolar apex, being taken at 100, that from side to side is 90, the axis is 600 ; ventral radius 45, dorsal 55. The section is slightly oval, the ventral face rather broader than the dorsal.

In a young specimen, 3 inches long, the proportions of the diameters are also 100 to 90 ; the axis is more excentric than in the older specimen, the ventral radius being only 30, the dorsal 70, the axis 650.

Phragmocone. Known in a crushed state by specimens from near Chippenham. The uncrushed phragmocone has a slightly elliptical section. D'Orbigny gives a very elliptical section (55 to 45). The characteristic angle, as given by D'Orbigny, is 16° 30'. I have no good specimen of this part of the fossil.

Locality. In the Oxford Clay of Wiltshire, Oxfordshire, Northamptonshire, Huntingdonshire. In the Kelloway Rock of Yorkshire. In the Oxford Clay at Vaches-Noires (Calvados), and Marquise, near Boulogne.

Observations. Specimens are often found invested with a sheath of white fibrous matter, externally rough and of granular aspect, within which the true shell is always smooth and shining.

I remark the following varieties:

1. Belemnites Owenii (Puzosianus). Pl. XXXI, fig. 76; Pl. XXXII, figs. 78, 79.

The guard is smooth and always compressed; the apicial furrow distinct or faint, never more than half the length of the axis; alveolar section elliptical. This is the ordinary form from the Oxford Clay of the Midland Counties. In one middle-aged specimen, corresponding to *B. attenuatus* of Mantell, lateral grooves extend all along the post-alveolar tract.

2. Belemnites Owenii (verrucosus). Pl. XXXI, fig. 77.

The surface is ornamented with small, raised, smooth puncta, and undulations composed of these united. The distribution of these may be seen on the ventral face (v'), the lateral (l'), and the dorsal (d'). On the first it will be noticed that the puncta disappear towards the apex, and diverge and disappear on the alveolar region. On the sides they show more of a tendency to gather in linear groups; on the back this concurrence of the puncta makes short undulated ridges, which grow larger, but more dispersed, on the alveolar region. The apex shows signs of very short plications. Only one specimen is known to me, found by Mr. J. E. Walker, at St. Neot's, with *Ammonites Duncani.* The reader may compare the curious granulation in this specimen with that on *B. infundibulum* (Pl. I, fig. 3), with that on specimens of *Belemnitella granulata,* and with the diverging ornaments on a Sepiostarium.

If further research should produce additional specimens, possibly there may be found reason to adopt a specific name for this fossil. But the surface-ornament being at present the only difference observed between this guard and ordinary specimens of *B. Owenii* of the same size, I prefer to mark it as a variety.

THE BELEMNITES OF THE OXFORD CLAY.

3. BELEMNITES OWENII (TORNATILIS). Pl. XXXII, fig. 80.

The guard is very smooth, less compressed than in the typical forms; more cylindrical, with a longer, deeper, and narrower ventral furrow (fig. 80). This furrow, indeed, occupies the greater part of the axial length of the guard; in middle-aged specimens an old specimen shows some trace of lateral flattenings. From the Kelloway Rock of Hackness and Scarborough. It is not unlikely that this may be found to deserve to be regarded as distinct specifically.

BELEMNITES STRIGOSUS, n. s. Pl. XXXII, fig. 81.

GUARD. Very long, slender, cylindro-conical, compressed, acuminated, smooth, with a distinct longitudinal furrow drawn from near the apex, on the ventral face, through two fifths of the length of the axis, and thence continued in a slighter depression towards the alveolar region.

Transverse sections of the guard show an oval contour, the sides flattened; the ventral face broader than the dorsal; in the alveolar region the dorsal part of the shell is much thicker than the ventral part.

Greatest length of the one specimen seen $7\frac{1}{4}$ inches, of which the axis is $6\frac{3}{4}$ inches; greatest diameter $\frac{45}{100}$ of an inch,

Proportions. The diameter at the alveolar apex from back to front ($\frac{4}{10}$ of an inch) being taken at 100, that from side to side is about 80, ventral radius about 40, dorsal radius about 60, axis 1600.

PHRAGMOCONE. Unknown. The alveolar section is nearly circular, the angle appears to be about 20°.

Locality. In the upper part of the Oxford Clay, in Cowley Field, near Oxford; one specimen, presented to the University Museum by W. B. Dawkins, M.A., the first Burdett-Coutts Scholar.

Observations. This remarkable fossil carries to extreme length the essential characters of the group of tornatile Belemnites, the cylindro-conical outlines, the slight compression, the apicial groove, and low angle of phragmocone. Having but one example to consider, I am unable to describe the variations due to age and accident, but it would be very agreeable to be furnished with evidence on these points. I have seen no foreign specimens corresponding with this species; but D'Orbigny's fig. 3, pl. xvi, somewhat resembles it. A thin white external layer appears on the specimen, not the fibrous layer noticed in *B. sulcatus* and *B. Owenii*.

BELEMNITES SPICULARIS, n. s.   Pl. XXXVIII, fig. 82.

GUARD.   Cylindrical (hastate when young), tapering evenly to a point, much compressed to an oval section, with a faint ventral groove drawn from the apex through two fifths of the length of the axis; a few striæ about the apex, especially on the dorsal aspect.

Transverse section oval, the ventral face broader than the dorsal.   Substance varied by bands of brown (sepia-tint) and honey-yellow spar.

Greatest length observed 10 inches, greatest diameter 1 inch.   Shortest specimen 1 inch long; it is of the form Pl. XXXIII, fig. 82 *l'*.

Proportions (old).   Taking the diameter at the aveolar apex at 100, the diameter from side to side is 90 +, the axis 1000; the excentricity of the axis variable, in some specimens small, in others the ventral radius = 40, the dorsal 60.

PHRAGMOCONE.   Incompletely known.   The section is elliptical, within a ring of the guard-fibres everywhere of nearly equal thickness; the phragmocone section more elliptical, therefore, than the section of the guard.   The angle in one of Lieut. Patterson's specimens appeared to be 18° at the apex, 15° in a more advanced part of the shell.   The apex of this phragmocone was placed at about one third of the diameter from the ventral margin.

Locality.   Eathie Burn, and Shandwick, on the coast of Cromarty: collected in great abundance and in excellent condition by Lieut. Patterson, who gave me much information as to the circumstances under which he obtained the specimens and the accompanying fossils.   He further assisted my researches by presenting to me a set of photographic representations of much interest.

The fossiliferous strata of the Mesozoic system on this coast have been usually described as Liassic, and on a first view of the shale and these Belemnitic fossils such an opinion might be readily adopted.   The Belemnite now in question has *analogy* to some of the long species of the Upper Lias, such as *B. tripartitus,* while the next to be mentioned seems to revive the memory of *B. longissimus* of Miller.   Their *affinity,* however, is with the long species of the Oxonian stage in the Oolitic system.   Among the accompanying fossils I observed in Lieut. Patterson's collection *Gryphæa dilatata,* large and small; *Perna; Avicula Braamburiensis; Pleurotomaria; Ammonites* resembling, if not identical with, *A. vertebralis,* Sow., *A. excavatus,* Sow., *A. flexicostatus,* Phil., *A. plicatilis,* Sow., *A. Gowerianus,* Sow., *A. biplex,* Sow; scales of *Lepidosteus;* cervical vertebræ of *Ichthyosaurus.*   The Belemnites form a bed in the shale.

Observations.   It is difficult to fix upon any definite characters by which to distinguish this Belemnite from *B. Owenii,* except the greater proportionate length of the axis and the faintness of the apici-ventral groove.   The slight striæ about the apex are only seen on one or two specimens.

BELEMNITES OBELISCUS, n. s.   Pl. XXXIII, fig. 83.

GUARD. Very long, almost uniformly tapering to a point, compressed, smooth, or with traces of longitudinal interrupted undulations. In some specimens a defined lateral flattening (Pl. XXXIII, l', l^3). No distinct apici-ventral groove.

Greatest length observed $9\frac{1}{10}$ inch ; greatest diameter in this specimen, just before the conical expansion, less than $\frac{1}{2}$ an inch. In shorter specimens, $6\frac{1}{2}$ inches long, the corresponding diameter is nearly the same ; in still smaller examples, $3\frac{1}{3}$ inches long, the diameter is $\frac{1}{8}$ of an inch. It seems as if two varieties exist, one much longer in proportion than the other.

Proportions. The normal diameter at the alveolar apex being taken at 100, the transverse diameter is 84 ; the axis in the longer variety 2000 and more, in the shorter 1500. The excentricity of the axis appears to be very small.

PHRAGMOCONE. I have only been able to observe the cross section, which is less elliptical than the sectional outline of the guard, the guard-fibres being longer on the back and front than on the sides. In this the fossil is analogous to some Liassic forms.

Locality. Eathie Burn and Shandwick, with the last-named species.

Observations. Not only do the unequal proportions of different specimens suggest the idea of a sexual distinction, but the whole group, compared with *B. spicularis,* leads to reflections of the same order. The guard is colour-banded, as is that of *B. spicularis.*

ON A GROUP OF BELEMNITES ALLIED TO *BELEMNITES EXCENTRICUS* OF BLAINVILLE.

Lister, in his 'Historia Anim. Angliæ,' pp. 226-228, has the following description of a Belemnite of this group :—" Titulus xxxi.—*Belemnites niger,* maximus, basi foratâ." Among the remarks on this species we find " Perfricatum cornu combustum aut quoddam bitumen olet." " In tota illa agri Eboracensis regione montosâ, qui *Blackmore* appellatur, præcipuè abundant ; item in rivulo juxta *Bugthorp,* et alibi reperti sunt." The Blackmore fossils belong to *B. abbreviatus;* a large fragment was above three inches in circumference. Bugthorp is on the Lias.

Lhwyd, in the ' Lithophylacium Britannicum,' notices Belemnites of this group, from the vicinity of Oxford, No. 1667 :—*Belemnites maximus oxyrrhynchus,* four inches in girth where largest. Cowley, Bullington, Marsham, Stansford, Garford, the localities mentioned, indicate the species to have been what Miller called *B. abbreviatus.*

Smith, in the 'Stratigraphical System,' p. 50, describes a Belemnite as elongate, rather four-sided, from Wotton Basset and Shippon, in the Coral Rag, p. 43, and another, quite similar, from the Kimmeridge Clay of North Wilts.

Miller described these forms as *B. abbreviatus ;* his followers have often assigned that name to a species from the Inferior Oolite.

Young and Bird, in their volume on the Yorkshire coast (ed. 1, 1822), notice a similar Belemnite, and give a figure (pl. xiv, fig. 4), and name it *B. excentralis*, describing it as found in the " Oolite, Upper (Speeton) Shale, and Chalk." This is incorrect, but, as will be seen, the large Speeton Belemnite belongs to the same natural group.

De Blainville, in 1827, describes and figures in his pl. iii, fig. 8, 8*a*, *B. excentricus*, from Vaches-Noires, remarking that Miller's *B. abbreviatus* much resembles it.

D'Orbigny revives Young and Bird's name for a species which he figures (pl. xvii); but in the text (p. 120) he makes no reference to those authors, and uses the name given by Blainville.

In the second edition of the first volume of my work on the ' Geology of Yorkshire' I restored to the great Belemnite of the Malton Oolite the name assigned by Miller, and mentioned the large Speeton Belemnite as *B. lateralis*. An undescribed form in the Kimmeridge Clay of Oxfordshire, and another in the Tealby beds of Lincolnshire, will complete this series of excentral Belemnites, as far as I know them.

BELEMNITES ABBREVIATUS, *Miller*.   Pls. XXXIV, XXXV, figs. 84—93.

> *Reference.*   *B. niger maximus*, Lister, ' Hist. Anim. Angliæ,' p. 226, 1678.
> *Belemnites maximus oxyrrhynchus*, Lhwyd. (No. 1667.) 1699.
> *B. excentralis* (in part), Young and Bird, ' Geology of the Yorkshire Coast,' pl. xiv, fig. 4, 1822.
> *B. abbreviatus*, Miller, ' Geol. Trans.,' 2nd series, vol. ii, pl. vii, figs. 9, 10, 1823.
> *B. excentricus*, Blainville, 'Mém. sur les Bélemn.,' p. 90, pl. iii, f. 8, 1827.
> *B. excentricus* (also called *excentralis*), D'Orbigny, ' Pal. Franç., Terr. Jur.,' p. 120, pl. xvii, 1842.

GUARD. Cylindrical; sides flattened or somewhat hollowed longitudinally; apex produced, compressed, sometimes incurved; ventral surface broader than the dorsal; a flattening near the apex, on the ventral surface.

Very old specimens have the apicial region much compressed, produced, and incurved; sides flattened by broad, shallow, longitudinal depressions, which continue over a part of the alveolar region, and are there gradually lost.

Young specimens are slightly hastate, very young ones distinctly so, with little trace of the lateral hollow.

Longitudinal sections show the axial line to be very excentric, especially so in the retral part of the guard, and in old specimens considerably curved.

Transverse sections present a somewhat four-sided outline, the ventral surface being struck to a flatter curve than the dorsal, and the sides flat or a little concave.

The length of a very large example is 8 inches; of another smaller, but extending farther along the phragmocone, 11 inches; the greatest diameter before the conical expansion of the sheath over the phragmocone $1\frac{1}{4}$ inch. The smallest which has occurred to me is little more than 1 inch long.

Proportions in full-sized specimens. Taking the dorso-ventral diameter at the alveolar apex at 100, the transverse diameter is about equal to it, the ventral radius is 32, the dorsal 68, the axis 250, justifying Miller's title of *B. abbreviatus.* In young specimens the axis = 300. Cross sections near the apex show a still greater excentricity, the axis curving towards the ventral surface.

PHRAGMOCONE. Conical, a little incurved towards the ventral line, with an almost perfectly circular section; sides inclined at an angle of 18°, except near the apex, where it is greater (above 20°). The septa are numerous, lie at right angles to the axis, with plain unwaved edges, and are penetrated by a marginal siphuncle.

Locality. In the Coralline Oolite and Calcareous Grit of Yorkshire at Malton, Scarborough, &c. In the same strata at Heddington, Cowley, Bullington, &c., near Oxford; near Calne, Weymouth, &c. In France, at Vaches-Noires and Marquise in Oxford Clay. D'Orbigny quotes it from near Moscow. In Oxford Clay, Cowley, near Oxford. In Kimmeridge Clay, Shotover.

Varieties. In progress from youth to age, this Belemnite experiences considerable changes, as may be inferred from what has been said in respect of the guard. Besides these ordinary changes of form and proportion, it appears desirable to distinguish two types of general shape, which occur in large specimens in some degree of relationship to the stages of the strata.

a. BELEMNITES ABBREVIATUS (OXYRHYNCHUS).

Large, *cylindroidal*, slightly bent, with incurved, produced, flattened apex. Viewed on the front or back, the sides are seen to contract rather suddenly from a cylindroid part to the apex in the architectural form known as ogee (fig. 84, ϕ); viewed sideways, the dorsal outline is continued in a convex form to or nearly to the apex, while the ventral outline becomes concave under the apex (Pl. XXXIV, fig. 84, l').

Locality. The Coralline Oolite of Malton, Oxford, and Wilts. An abnormal specimen, which places these characters in a strong light, is presented (Pl. XXXV, fig. 86) from near Oxford.

β. BELEMNITES ABBREVIATUS (EXCENTRICUS).

Large, conoidal, with sides almost straight, converging through the whole post-alveolar

19

space of the guard, the dorso-ventral diameter being in that part much greater than the transverse diameter. The general figure is that of *Belemnites explanatus* (Pl. XXXVI, fig. 96).

Locality. The Oxford Clay, upper part, at Cowley, near Oxford. The Calcareous Grit and Coralline Oolite near Oxford.

This variety agrees well with the description and figure of Blainville (' Mém. sur les Bélemnites '). D'Orbigny makes the phragmocone section to be more elliptical than it usually is, and the axis less excentric than usual; there is also something about the outlines not as we commonly see them.

REMARKS ON SPECIMENS OF *BELEMNITES ABBREVIATUS*, VAR. *EXCENTRICUS*, IN THE CABINET OF MR. WETHERELL.

The axial line of the guard is in many instances excavated into a canal which grows narrower towards the apex. This is especially the case in specimens obtained from the Drift of Finchley, near London, from which a great variety of fossils of the Oxford Clay and other strata lying to the northward has been obtained by Mr. Wetherell. In the large collection of that gentleman are very many excellent examples of this structure, and by careful study of them in comparison with other undisturbed specimens in the Oxford Clay, Calcareous Grit, and Oxford Oolite, we arrive at a clear view of a very curious subject, of which, at first sight, it might seem difficult to form a correct opinion.

Fig. 88, Pl. XXXV, represents the surface produced by a splitting fracture through a Belemnite in Mr. Wetherell's collection. Fractures of this kind are not infrequent in nature, and are easily produced by intention. The surface thus presented is usually flat and smooth in the ventral portion, as if a natural fissure existed there, but commonly uneven and more or less hackly in the dorsal portion. The hollow left by the spherule is sometimes traceable at the apex of the alveolar cavity, the phragmocone being generally absent in the specimens under consideration, but not seldom the alveolar cavity contracts gradually to the canal without any distinct enlargement at the alveolar apex.

In the figure referred to the canal is seen to contract gradually until it finally dies out before reaching the apex. Examined with microscopic powers of 10 and upwards, the canal is found to be crossed by many ridges at nearly equal intervals, so as to suggest the appearance of an annulated or half-chambered canal, in continuation with the cavity or the spherule of the phragmocone (see fig. 91). The seeming septa of this canal are found by more careful research to be the truncated edges of the successive laminæ of the guard (see figs. *m'*, *m''*), each conspicuous lamina giving origin to one septum. This appears quite certain under the lowest power of a good achromatic microscope, which discloses, moreover, that the laminæ thus referred to appear to be often composed of two or three thinner layers, some dark, others paler, and probably more nacreous in substance.

Following the canal till it closes, the laminæ are seen to lose their truncations, and to acquire the complete curvature.

After careful study of many specimens, no doubt remains in my mind that the canal has been produced by the removal of the apices or terminal parts of the interior laminæ of the guard. This process began at the alveolar cavity; it happened during life, and was occasioned by decay and absorption of the apices in the earlier stages of life.

That these special parts might be of somewhat different composition from the other parts of the laminæ is suggested by some other cases in which terminal porosity and an axial canal have been noticed; and it is quite in agreement with two other circumstances to be observed in these fossils. First, it is to be remarked that the alveolar cavity in these Belemnites often appears marked by the undulated anterior edges of the laminæ of the guard, which terminate in this cavity (see figs. 88, 91, 92), and show white, thin, sparry plates, in consequence of the removal of parts of the laminæ. And again, some of the specimens show a curious appearance of a second canal going from the alveolar cavity (figs. 90 and 92) near its apex. This, being carefully studied, is found to be occasioned by the removal of some of the laminæ of the sheath for a certain space inwards from the alveolar cavity, leaving a kind of slit where the removal has happened.

In later life the deposited sheaths were, in general, not removed by decay or absorption (see figs. 87, 88).

ON THE BELEMNITES OF THE KIMMERIDGE CLAY.

After diligent search in this clay near Oxford, where it is about 100 feet thick, and is pretty well exposed in brickyards and in quarries of the Coralline Oolite, and after a careful search in the escarpment of Portland, I find, speaking generally, a remarkable accordance between its Belemnites and those of the Oxford Clay known as *B. abbreviatus excentricus*, *B. Owenii*, and *B. hastatus*. This analogy was, perhaps, to be expected, inasmuch as Ammonites of the groups of *A. vertebralis* and *A. biplex* occur in both clays.

Taking first the specimens allied to *B. abbreviatus excentricus*, it would, I think, be difficult to assign characters of sufficient weight to claim a specific distinction, though in old specimens the ventral surface is more flattened towards the apex, and in young specimens the whole of the guard is depressed behind the alveolar region. In this respect the young forms closely resemble those of *B. Souichei* (D'Orbigny, 'Ter. Jur.,' pl. xxii, figs. 4—8), which was found in beds referred to the Portland series at Hauvringhen (Pas de Calais), and at the Tour de Croi, near Boulogne. These forms differ from those of the same age from the Oxford Clay and Coralline Oolite.

Next, we may consider the longer forms, like *B. Owenii*, of the Oxford Clay. Of these some appear to me quite indistinguishable from their analogues in the older deposit ; they occur of equal magnitude with them, but not in equal abundance, in the upper part of the Kimmeridge Clay of Shotover Hill, where it was cut through by the railway. One extremely lengthened variety of this Belemnite occurs at Shotover, reminding us of *B. spicularis* from the shore of Cromarty.

Besides these is a young depressed Belemnite much like the young *B. hastatus* of the Oxford Clay ; these occur near Oxford and in Portland Isle.

BELEMNITES EXPLANATUS, n. s.   Pl. XXXVI, figs. 94, 95, 96.

GUARD (old).   Conoidal, tapering gradually to a rather compressed apex ; sides more or less broadly channelled ; ventral aspect flattened and somewhat expanded, becoming concave towards the apex (a few dorsal striæ about the apex are sometimes seen).

Dimensions.   Axis about 3 inches ; diameter at the alveolar apex 0 85 inch.

(Young.)   Depressed, smooth, flattened on the ventral aspect, and hollowed, or marked by a narrow groove towards the apex, which is slightly curved ; sides more or less marked by a shallow continuous furrow (a very young form is almost fusiform).

Dimensions.   Axis in the smallest about 1 inch, with a diameter of 0·25 inch.

Proportions.   Axis in young specimens 400—450. The diameters at the alveolar apex 100 from front to back, 115 from side to side. In old specimens the axis is about 350, the transverse diameter 107, the dorsal radius 64, the ventral 36.

Locality.   In the upper part of the Kimmeridge Clay of Waterstock, near Thame. Specimens of different ages — young (not middle-aged) and full-grown—have been presented to the Oxford Museum by Mrs. Ashworth. In the upper part of the same clay at Hartwell, near Aylesbury, with *Cardium inæquistriatum, Astarte Hartwelliana,* and *Ammonites biplex.* In the Kimmeridge Clay, upper part, where cut through in the railway-tunnel, at Wheatley, near Oxford.

Observations.   On many accounts this form of Belemnite is of interest in the study of the series to which it belongs. On the one hand its resemblance to the older type of *B. abbreviatus* (*excentricus*) of the Oxford Clay and Oolite, and on the other to that of Speeton, in Yorkshire (*B. lateralis*), is such as to offer a most instructive example for study, in relation to the derivation of successive specific forms by hereditary trans-mission with modification. But this must be considered hereafter.

EXPLANATION OF PLATE XXVIII.

Fig

67. BELEMNITES HASTATUS.

Three views of a large specimen from the upper part of the Oxford Clay in Cowley Field, near Oxford; in the Cabinet of Mr. Parker. Fig. 67, l, lateral view, showing the rather faint oblique groove; v, the ventral aspect, with its deep characteristic furrow, suddenly followed by a shallower channel; d, dorsal aspect; s', the cross section of the alveolar cavity; s'', section across the expanded part of the guard.

68. Two views of a specimen from the Oxford Clay of Northamptonshire, presented to the Oxford Museum by Mr. Leeds, B.A., of Exeter College.

l'. Lateral view, showing the oblique groove; v', ventral aspect, showing the deep mesial groove expanding retrally; s', the cross section of the alveolar cavity; s'', cross section of guard in the narrowest part; s''', cross section in the widest part; s'''', cross section near the apex; ϕ, a compressed phragmocone, partly covered by the guard.

69. Young specimens from the Oxford Clay of St. Ives, presented to the Oxford Museum by Mr. J. F. Walker, of Sidney Sussex College, Cambridge.

l^2. Lateral view, showing a recurved apex; v^2, ventral aspect of the same; s^1, cross section of alveolar cavity; s^2, cross section of enlarged guard; v^3, ventral aspect of a younger individual; l^4 and v^4, views of a still younger shell; v^5, l^5, views of a shorter and more bulbous example; v^6, v^7, views of the bulbous variety, with interrupted ventral groove.

70. μ. Deformity. The specimen is from the Oxford Clay of Cowley Field, near Oxford. Cabinet of Mr. Parker.

PL. XXVIII.

Fig. 68.　　　Fig. 67.　　　　　　Fig. 68.

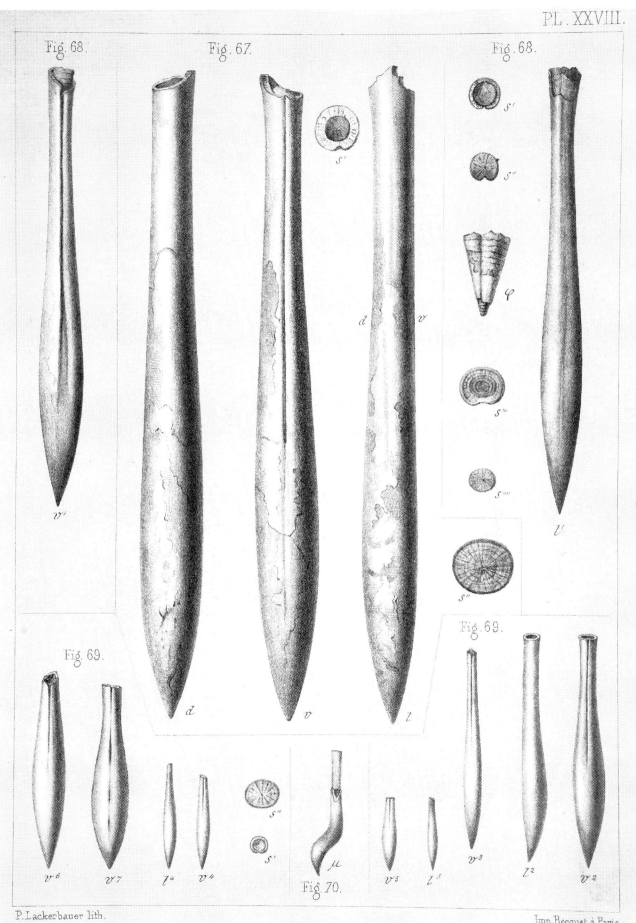

Fig. 69.

Fig. 69.

Fig 70.

P.Lackerbauer lith.　　　　　　　　　　Imp.Becquet à Paris.

EXPLANATION OF PLATE XXIX.

Fig.

71. BELEMNITES SULCATUS. Specimens from the Oxford Clay at Summertown, near Oxford.

> v. Ventral aspect of a full-sized specimen ; the groove somewhat less distinctly marked than in other cases, and slightly interrupted.
>
> v'. A younger specimen, showing a more marked change of depth in the ventral furrow and its continuation over the alveolar region ; s', cross section of the alveolar region of the guard, depressed.
>
> s'. Longitudinal section, from back to front, showing the phragmocone *in situ*, its apical spherule, and a short lanceolate canal, formed by the decomposition of laminæ, as shown in A.
>
> s''. Another longitudinal section showing very similar facts ; the canal somewhat more extended in A''.
>
> s. Cross section of the guard, showing the inflection of the laminæ to form the ventral groove, and lacunæ of a remarkable kind.

72. Young specimens from Summertown, near Oxford ; *l*, lateral ; *v*, ventral aspect ; *s'*, alveolar cross section, nearly round ; *s''*, post alveolar section, nearly round ; *m*, magnified surface, the shell dotted with granules, and covered by a partially fibrous layer. Such a layer occurs on some Australian and on some Indian Belemnites.

73. Youngest examples which occur near Oxford, at Cowley, and Long Marston ; *v*, ventral aspect, the groove distinct on the alveolar region ; *v'*, the still younger shell ; *s*, alveolar cross section.

Fig. 71.

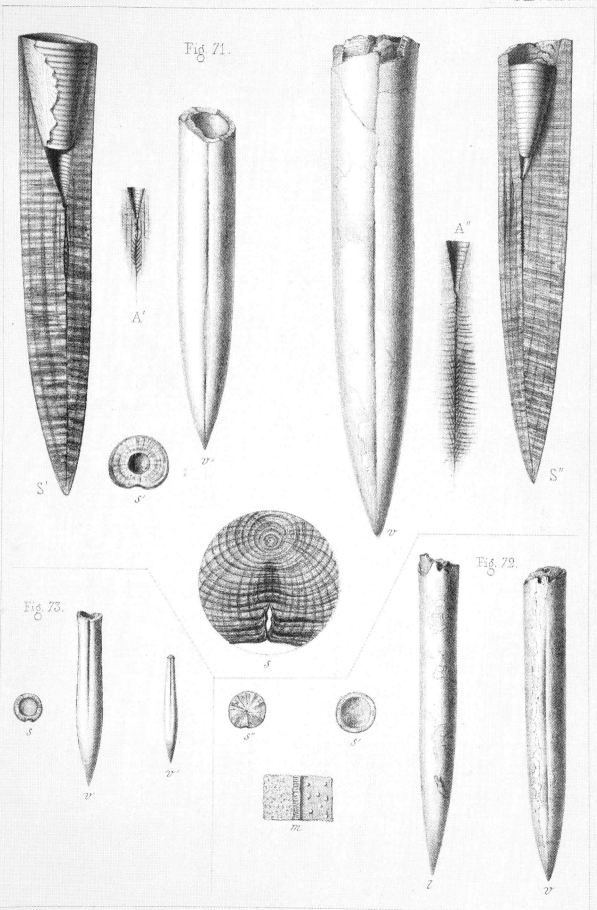

A'

S'

s

v'

A"

S"

v

Fig. 73.

s

Fig. 72.

v'

v

s

s"

s'

m

l

v

P.Lackerbauer lith.

Imp.Becquet à Paris.

EXPLANATION OF PLATE XXX.

Fig.

74. BELEMNITES SULCATUS. Specimens of full size from the Oxford Clay of St. Neot's; presented by Mr. J. F. Walker, of Sidney Sussex College, Cambridge; *l*, lateral view, showing a very slight flattening; *v*, ventral aspect, showing the groove widening and growing shallow over the alveolar region, partially interrupted toward the apex, with striation parallel to the groove; *s*, cross section of guard, slightly depressed.

75. Specimens from Weymouth. *v′*, ventral surface, showing the groove growing wider and shallower both toward the apex and over the alveolus; *d′*, dorsal aspect; *s′*, cross section of guard nearly round Fig. 75, *l″*, lateral view of another specimen, showing a slight flattening; *v″*, ventral aspect, the groove widening over the alveolar and apicial regions, with striæ of decomposition parallel to the groove; *s″*, alveolar cross section; *s‴*, cross section of guard.

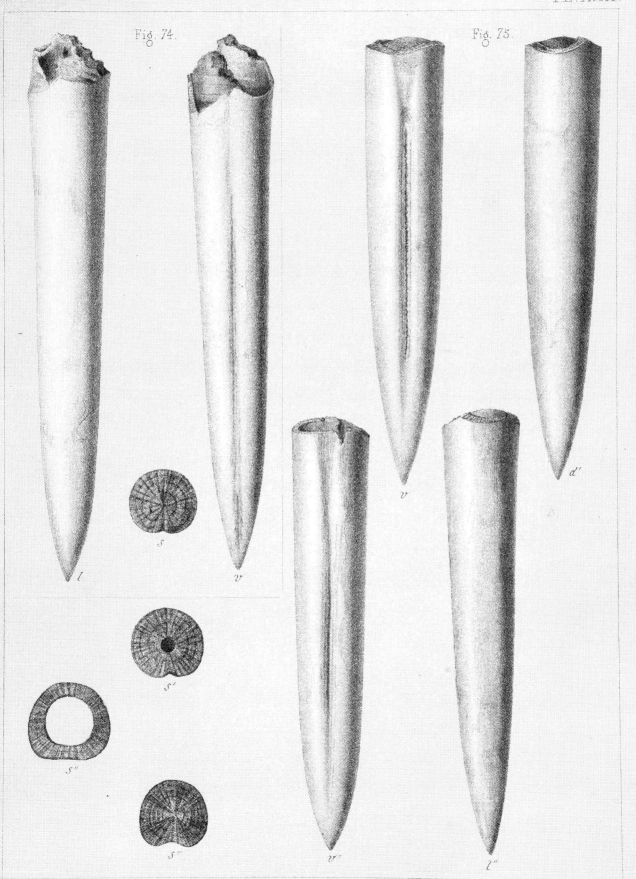

Fig. 74.

Fig. 75.

EXPLANATION OF PLATE XXXI.

Fig.

76. BELEMNITES OWENII, var. PUZOSIANUS. Specimens from the Oxford Clay of St. Neot's, presented by Mr. J. F. Walker.

l, full-sized specimen seen laterally; *v'*, middle-sized individual, showing the apicial groove distinct; *v''*, younger individual seen ventrally, with its apicial groove; *l''*, side view of the same; *v'''*, still younger example seen ventrally; and *l'''*, seen laterally.

77. BELEMNITES OWENII, var. VERRUCOSUS. From St. Neot's, presented by Mr. J. F. Walker.

v', seen ventrally; *l'*, laterally; *d'*, dorsally; *s'*, cross section in the alveolar region.

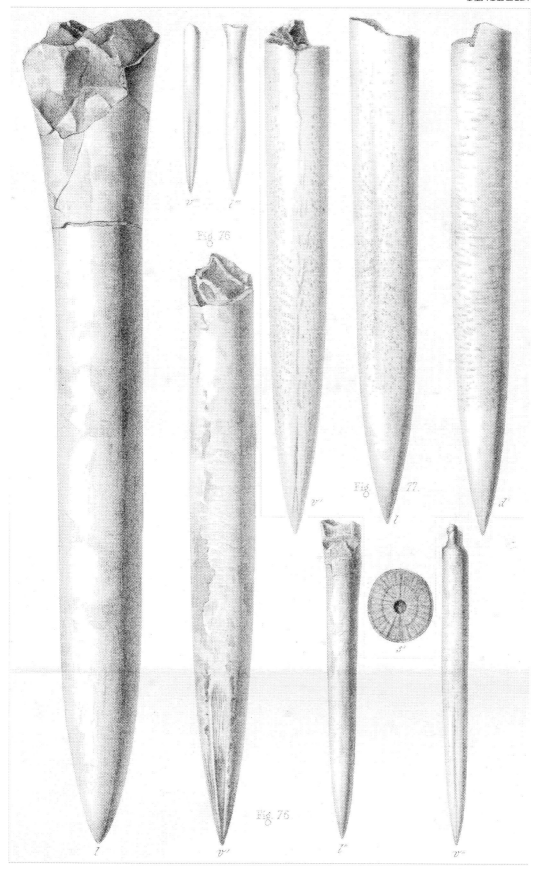

Fig. 76

Fig. 77.

Fig. 76

EXPLANATION OF PLATE XXXII.

Fig.

78. BELEMNITES OWENII, var. PUZOSIANUS. From St. Neot's, presented by Mr. Walker; *v*, ventral face, with strongly marked apicial groove reaching to the point; *l*, lateral view. The cross section is oval.

79. BELEMNITES OWENII, var. PUZOSIANUS. From St. Neot's; *v*, ventral aspect, showing an almost evanescent apicial depression; *d*, dorsal aspect; *s*, the cross section of the alveolar region.

80. BELEMNITES OWENII, var. TORNATILIS. From the Kelloway Rock of Hackness, near Scarborough; *v*, the ventral surface expanding anteriorly; cross section oval.

81. BELEMNITES PORRECTUS, n. s. From the Oxford Clay at Summertown, presented by Mr. Dawkins; *v*, ventral aspect, showing a strong, sharply cut apicial furrow, and its anterior extension; cross section oval.

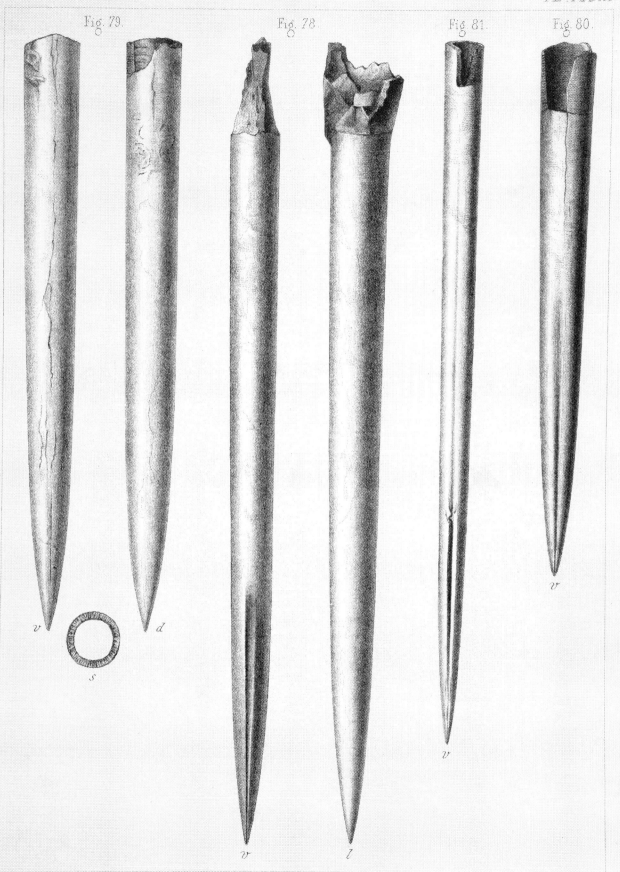

Fig. 79.   Fig. 78.   Fig 81.   Fig. 80.

P. Lackerbauer lith.   Imp Becquet à Paris.

EXPLANATION OF PLATE XXXIII.

Fig.

82. BELEMNITES SPICULARIS, n. s.

Specimens from Shandwick and Eathie, on the coast of Cromarty, collected by Lieut. Patterson ; *v*, ventral aspect of one of the larger specimens, showing the apicial groove ; *l*, lateral view, showing the long side flattening, and a trace of short apicial grooves ; *d*, the dorsal aspect, marked by some small striæ ; *l'*, one of the youngest specimens in Lieut. Patterson's series ; *s*, cross section of guard at the alveolar apex ; *s'*, section of a smaller specimen toward the apex ; *s''*, still nearer the point.

83. BELEMNITES OBELISCUS, n. s.

Specimens collected by Lieut. Patterson at Shandwick and Eathie ; *l'*, one of the longest examples, seen sideways; *l''*, one somewhat smaller, *l'''*, still smaller, and *l*^{iv}, one of the smallest observed ; *v'''*, ventral aspect of a small specimen, and *v*^{iv}, of a still smaller ; *s'*, alveolar cross section ; *s''*, post-alveolar section across the guard; *s'''*, farther backward in the guard.

Fig. 82.

Fig. 83.

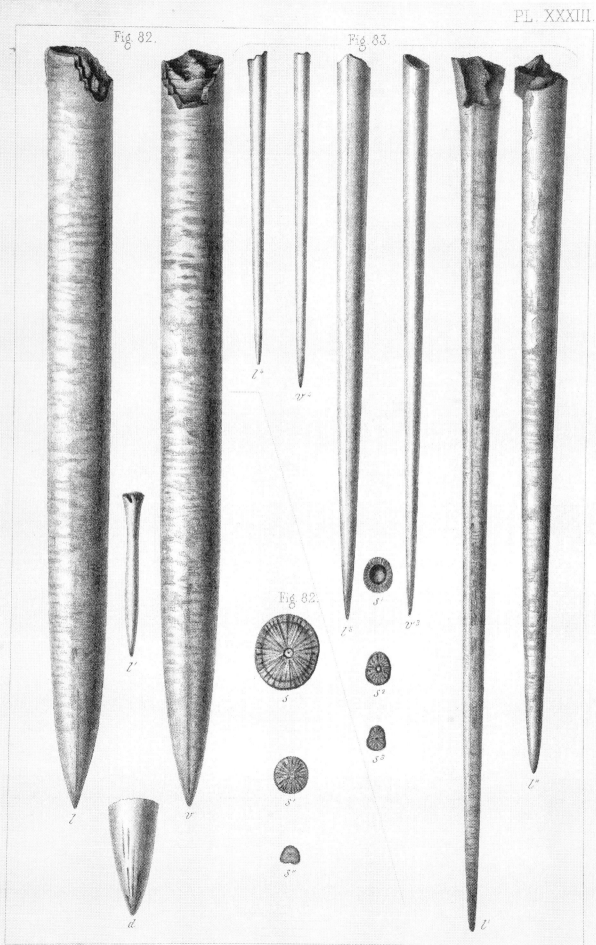

Fig. 82.

P. Lackerbauer lith.

Imp. Becquet à Paris.

EXPLANATION OF PLATE XXXIV.

Fig.

84. Large specimens of BELEMNITES ABBREVIATUS, from the Coralline Oolite of Yorkshire and Wiltshire.

l'. Side view, showing a broad lateral depression, and the incurved apex; *v'*, ventral aspect, showing a flattening near the apex. The specimen is from Malton, in Yorkshire.

φ. Showing the phragmocone *in sitú*, and the numerous septa, at right angles to the axis. The specimen is from the Calcareous Grit of Seend, in Wiltshire. The drawing was made by Miss Anne Cunnington.

s'. One of the septa seen axially; *s''*, cross section of the alveolar region; *s'''*, cross section behind the alveolus; *s''''*, section near the apex, to show the compression.

85. Younger individuals from Heddington, near Oxford, and Malton, in Yorkshire.

l''. Middle-aged specimen seen laterally; *s''*, alveolar section of the same.

v''. Specimen seen ventrally, with distinct apicial depression; *v'''*, smaller example, a little hastate; *v*^{iv}, the youngest observed.

l''', A very young specimen seen sideways, a little hastate.

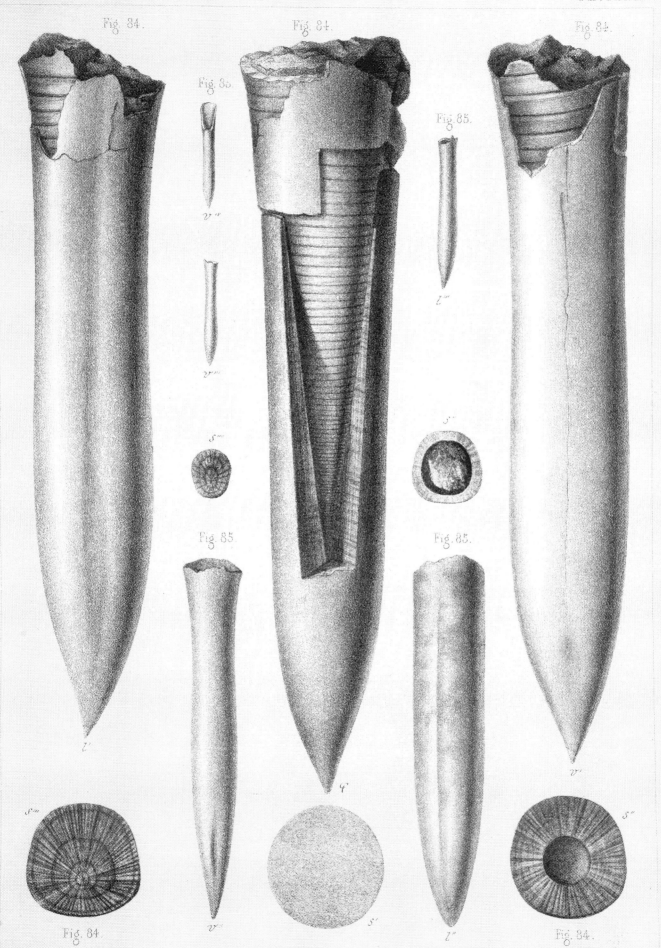

Fig. 84.　　　　　Fig. 84.　　　　　Fig. 84.

Fig. 85.

Fig. 85.

Fig. 85.　　　　　Fig. 85.

Fig. 84.　　　　　　　　　　　Fig. 84.

P. Lackerbauer lith.　　　　　　　　　　　Imp. Becquet à Paris.

EXPLANATION OF PLATE XXXV.

Fig.

86. A specimen of BELEMNITES ABBREVIATUS (seen laterally), compressed, and unusually bent at the apex; Heddington.

87. Section of BELEMNITES ABBREVIATUS, with the phragmocone *in situ*. The bending of the axial line of the guard is not often so remarkable as in this case, even in old specimens; it is characteristic of full growth.

88. A specimen with less curvature at the point, and less flexure of the axial line. From the Drift of Finchley, in Mr. Wetherell's collection.

89. Natural section to show the decay of the axial laminæ at their apex; Bullington, near Oxford.

The following figures are taken from specimens in the Collection of Mr. Wetherell, from the Drift at Finchley.

90. Section showing the formation of an axial canal and a vicinal fissure on the ventral side.

91. Section of another specimen.

92. Other sections, in which a small collateral ventral slit appears in the laminæ.

93. Section in which the canal appears interrupted.

m', m'', m''', m^{iv}. Magnified views of the laminæ in relation to the canal.

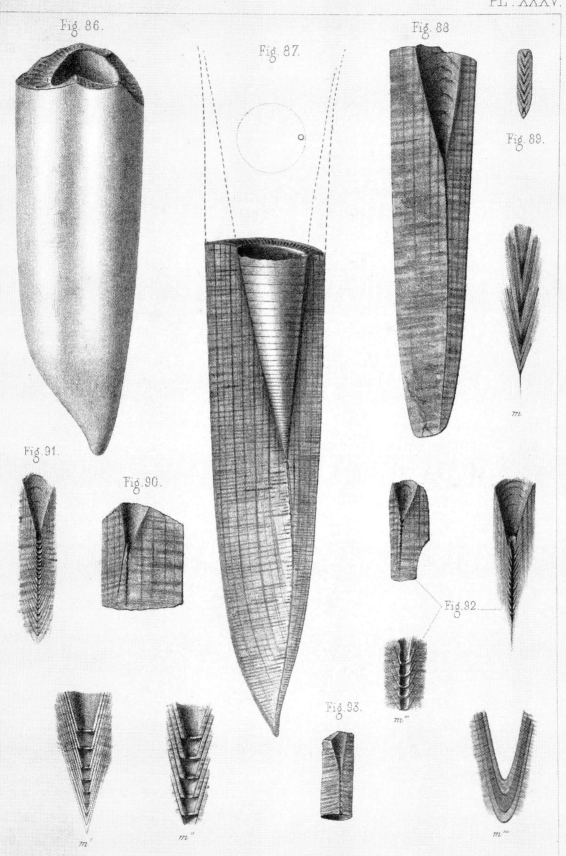

Fig. 86.

Fig. 87.

Fig. 88

Fig. 89.

m.

Fig. 91.

Fig. 90.

Fig. 92.

m'''

Fig. 93.

m'

m''

m''''

P. Lackerbauer lith.

Imp. Becquet à Paris.

EXPLANATION OF PLATE XXXVI.

Fig.

94—96. BELEMNITES EXPLANATUS, n. s.

94. Young specimens from Aylesbury, in the upper part of the Kimmeridge Clay.

v', v'', v'''.   Aspect of the ventral face, showing the flattening and slight furrow toward the apex.

l'''.   Lateral aspect, showing the longitudinal depression.

s'.   Section across the alveolar region; s'', section further back on the guard.

95. Specimens more advanced in growth, from Waterstock and the Railway-cutting in Shotover Hill, in upper part of Kimmeridge Clay.

l^{iv}.   Lateral view, showing the longitudinal groove; v^{iv}, ventral aspect of the same, showing the apicial flattening; v^{v}, the same view of a somewhat larger individual.

s'''.   Section across the alveolar cavity; s^{iv}, across the same cavity near its apex.

96. Full-grown individual from Waterstock, upper part of Kimmeridge Clay.

v^{vi}.   Seen on the ventral aspect; the apicial depression wide.

l^{vi}.   The same seen sideways; the lateral depression very distinct, the apex somewhat bent downward.

d^{vi}.   The same seen dorsally, where no trace of furrow appears.

s^{v}.   Section across alveolar cavity; s^{vi}, section behind the alveolar apex; a, the apex seen axially, to show its compression.

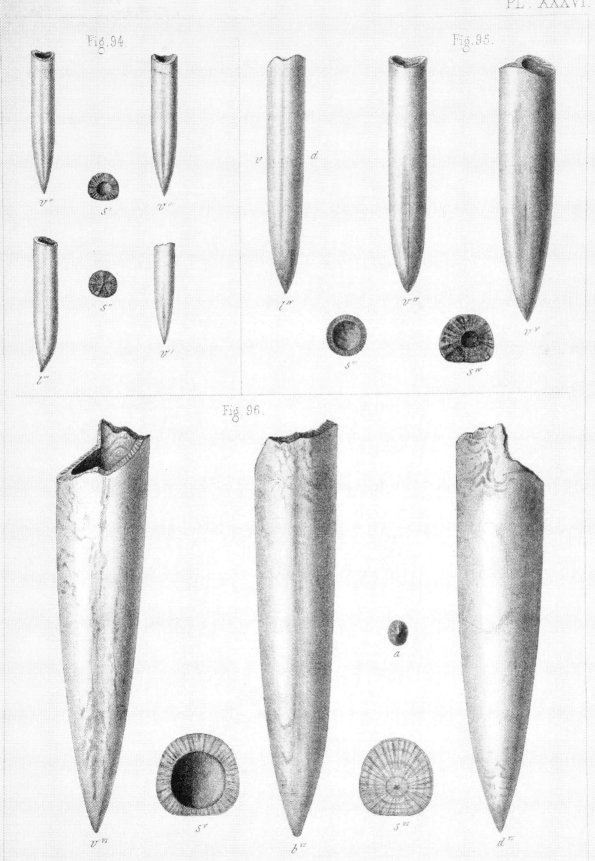

PL. XXXVI.

Fig. 94.

Fig. 95.

Fig. 96.

Imp. Becquet à Paris.

Printed in the United States
By Bookmasters